SILENT FREEDOM
A Memoir of Service with the 101st
Airborne Division (Air Assault) in Iraq

Aurea C. Franklin

SILENT FREEDOM

A Memoir of Service with the 101st
Airborne Division (Air Assault) in Iraq

Aurea C. Franklin

Aurea Press
WOODBRIDGE, VIRGINIA

© Aurea Press, 2021

All rights reserved. Except for brief quotations in a review, this book, or parts thereof, must not be reproduced in any form without permission in writing from the author. For information, address silentfreedom101st@gmail.com.

First published 2021

ISBN: 978-1-7375086-0-1

EPUB ISBN: 978-1-7375086-1-8

Printed in the United States of America

ACKNOWLEDGMENTS

I dedicate this book to the people I love. To my son, Christopher John Franklin, who inspired me to stay alive and return home safely. His sacrifice and continued bravery—when at such a young age he would watch the war on TV during Operation Enduring Freedom (OEF) and Operation Iraqi Freedom (OIF) hoping to catch just a glimpse of me—touched me to my core as his mother. To Krishna Jeyakumar, my daughter-in-law who met my son while they both attended George Mason University in Fairfax, Virginia, so many years after my return from Iraq. They married twice; I witnessed both wedding receptions—first in the United States, and then a wonderful, grand wedding in India a few months later. I dedicate this book to my family in India; thank you for everything and your continued moral support. To my stepchildren—Cliff Franklin II, his wife Jodie, and their son Evan, and to Patrick H. Franklin and his wife Abi—they were there for me before, during, and after my divorce from their father, and continue to be my family. To my family and friends in Honolulu, Hawaii: Mom and Dad, Dolly, Lani, Creighton, and Shane. Thank you for everything. I love you so much. To my family and friends in Manchester, Missouri, whose guidance I always treasured. I treated them as they treated me—like their own. To my parents who showed me all their love; may they rest in Heaven. To the rest of my family: to Emmy, Armand, Rose, Rosgil, and all their siblings, to my nephews and nieces, and to the grands in the Philippines and all over the world; they mean so much to me, and I to them, because at the end of the day, we are still a family. To my co-workers for their encouragement during the writing of this book. To Dr. Linda Williams, who helped make this book a reality. And to the men and women serving our country, the United States of America—thank you for your Heroism and Valor.

TABLE OF CONTENTS

	Acknowledgments	v
Part I	**Destination: Mosul**	1
	My Silent Freedom Is All That I Have	4
	The House of Abraham	11
	The Royal Tomb	12
	Fourth of July	14
	The War Zone	21
	Challenges	43
	CSH Redeployment	44
	Mosul, Iraq	47
	Bravo Seven	50
	Special Leave Accrual (SLA)	51
	Big Day	52
	Motivation	53
	Change of Command	54
	Battle Buddy	55
	Burn Pits	55
	Promotion Board Panel	56
	Shocking News	57
	Emergency Leave	57
	VIP Visit	61
	Freedom Goes to the United States	62
	Philanthropic Mission	62
	Stormy Weather	63
	Nineveh	63
	Blinding Dust Storm	64
	RPG Attack	65
	A Sad Day	66

Table of Contents

Memories	67
Dahuk, Iraq R&R	67
Ramadan	69
Qayyarah West (Q-WEST)	69
CH-47 Chinook Casualties	70
Challenging Time Off	71
CSM Wilson	73
The Capture of Saddam Hussein	75
Christmas in Iraq	77
Convoy to Freedom	79
Conclusion	83

PART II Second Deployment. Destination: Tikrit, Iraq 85

OIF-2	87
101st Airborne Division, Air Assault	91
101st Rakkasans Transfers Authority to Strykers	93
Change of Command	94
Change in Jobs	95
Wings of Destiny	97
Containerized Housing Unit (chu)	104
Other Facilities	108
MWR Facilities	108
Dust Storm	110
Flood in Tikrit	111
Bed of Oil	112
The Reason We Are at War	113
Live-Fire Training	116
Marksmanship	117
Increasing Insurgency	118
Lethal 101st CAB	119
OIF-2 R&R	121
Christmas at COB Speicher	122

Table of Contents

	USO	124
	FRIES Training	126
	Rocket Attack	127
	More Casualties	127
	Easter Sunday	128
	The Easter Celebration at the Airfield	130
	An Unforgettable Event	132
	PT and Aerobics after Lent	133
	Preparing for the PT Test	134
	Drill Sergeant McCroskey	136
	Silent Freedom at Its Best	137
	Drill Sergeant's Joker Grin	138
	Asian Pacific Month	140
	Hellcat Farewell BBQ Bash	141
	Preparation for Redeployment	142
	Change of Command Ceremony	143
	Redeployment First Sergeant	144
	Welcome Home	146
	Time Off	146
Part III	**Back to Iraq. Destination: The Zoo**	**149**
	Third Tour to OIF	151
	Reminiscence of the Past	155
	After Retirement from the Army	159
	Fruits of Labor	162
	Job Training	163
	Up, Up, and Away	164
	Hardest Tour	166
	Transitory Formation	169
	Work Challenge	171
	Driving Lessons	173
	The Biting Point	175
	Masterwork	178

Table of Contents

Camp Striker	180
The Road to Camp Striker	181
Linguist or Translator	189
Camp Victory	191
Thanksgiving at Oasis, 2007	194
Walking Buddies	199
Palace on the Hilltop	200
Olive Trees	201
Special Forces Group's Compound	203
Day and Night	204
SFG Fancy Dining Facility	205
Bootleg DVDs	206
Big Bird at the SFG DFAC	208
Farewell Party	211
Sad News	211
The Snowfall	213
R&R Time	214
Three Musketeers	216
Luis Returns from R&R	219
Rosemary's Disappearance	221
New HR Supervisor	222
The Company Spirit	223
Mail Runs	225
Yellow Zone	227
Unforeseen Dust Storm	230
Halloween in Iraq	233
Thanksgiving at the Zoo	239
Christmas at the Zoo	242
Reindeers in the DFAC	244
Commanding General's Visit to the DFAC	245
Christmas Mass at Camp Striker	248
The Birth of Jesus	249
Camp Victory Night	254

Table of Contents

	No One Was More Cheerful than Lydia	256
	Meeting with the Air Force	258
	Salsa Night	259
	Rumors	260
	The Surge	261
	Change of Contract	266
PART IV	**The Fertile Soil of Iraq**	**273**
	Aerial View	275
	On the Way to Camp Echo	278
	Kuwait Liberation Medal	282
	Camp Echo	284
	Samarra	289
	Back to Baghdad	293
	Dining Out at North End Pizza	295
	Remarkable Luau Site	300
	Arrival of the Hawaiian Decorations	303
	Farewell Luau	305
	A Bobcat and a Fox	309
	Heartbreaking News	311
	Paul's Visit and Tour of the IZ	312
	Camp Slayer	314
	Victory Over America Palace	315
	Ruins from the Bomb Attack	316
	Vainglory	317
	Perfume Palace	318
	More Ruins	320
	Mini Mosque	324
	Flintstone Cave	326
	Jack and Jill	327
	The PX	329
	Forbidden Intimacy	331
	Troop Withdrawal	336

Table of Contents

Revisiting the Park	338
The Musketeers to My Rescue	340
SFG Complex	345
The Kiss	346
Photo-ops with Lydia in Al-Faw Palace	348
The Bunkers	349
Al-Faw Palace's Jade Grand Entrance Door	353
Chandelier Madness in the Al-Faw Palace	354
Going Fishing for Saddam Bass	358
Unexpected Encounter in the Palace	359
My Redeployment	364
Afterword	367

PART I
DESTINATION: MOSUL

Rendezvous with Destiny

I had been trying to catch a ground assault convoy to northern Mosul to rejoin my unit. It was the only way to get there. I had the choice to repatriate with the 86th Combat Support Hospital (CSH), but my silent freedom told me to go to Mosul. *Silent freedom* is having a choice or choices on what to say or do, freely; it is free will. With silent freedom, I am free to think what I want to think and do what my heart and mind dictate, freely. My heart wants me to go to Mosul and join my unit. However, my commander gave me another choice; that is, I could retrograde with the CSH. I replied with due respect that maybe he is thinking of another soldier. Because I am not about to leave my fellow soldiers at war. We are in this together and I am going to Mosul. I have finished my mission with the CSH, and I want to go further north to Mosul. I was glad I made that choice and happy to be reunited with my family in the 101st Airborne Division, Air Assault (AASLT). Mosul is one of the historic centers of the Assyrians and their churches: the Chaldean Catholic Church, the Syriac Orthodox Church, and the Assyrian Church of the East, containing the tombs of several Old Testament prophets, such as Jonah, which I had the privilege to visit during the conflict. Now it had become a hotbed of conflict, causing our heroes to fight every single day. Freedom isn't free; it costs lives to be free, which I know very well to be true, because I had lost friends in this war. But we were there to win. And the price of winning brings me back to the words of my father.

I remember him saying over and over, "Fight for what is right. Always believe in yourself. When you dream, dream big and make that dream come true! And remember, it is a moral imperative that you do not discriminate. Treat others the way you want to be treated. If you would like people to treat you with respect, then you should ensure that you treat them with respect in return. It is all part of winning."

And I remember answering him every time he would finish, "Yes, father. Yes, father." And I would catch a gleam of pride in his eyes—and in my mother's eyes, too. I always strived to be someone my parents could be proud of—a good child. And I believe in what my father said. Some of my dreams did come true, and I am still working on my other dreams. I am grateful for my silent freedom.

My Silent Freedom Is All That I Have

When Sol and I got married, the whole village came out to witness our ceremony. I was introduced to Sol through a fellow instructor at the university. Her husband and Sol worked together in the Kwajalein Atoll in the Marshall Islands, at the Ronald Reagan Ballistic Missile Defense Test Site. They worked on various intelligence support systems, such as missile launchers, radar systems, tracking cameras, and many more, spread across many islands. Anyway, Sol was told about me at the time, and he wrote me a long letter stating that it was a pleasure to meet me, even if just in a letter. I cherished his first letter and hid it in my footlocker.

Sol finally decided to take a leave of absence from work to meet me and my loved ones. My aunt and uncle were the first ones to fall in love with Sol; they treated him like a son. My parents, especially my father, retained doubts about Sol's intentions—which I totally respected. My father had always been a disciplinarian; this attribute of his helped me survive the rigors of Army training, even in combat.

At our wedding ceremony, we had a dozen godparents, including my cousin who was a judge, and my fellow university instructors. I was wearing a light rose lace midi wedding dress, and my heels were just right—not sinking into the ground. The left side of my hair was adorned with white flowers. My hairdresser was my cousin and she said that my wedding would be remembered in our town for a long, long time. My niece Emmy was the maid of honor and her boyfriend Armand was the best man. Sol was wearing a *Barong Tagalog*, which is commonly worn as formal or semi-formal attire in Filipino culture and is worn untucked over an undershirt with belted trousers and dress shoes.

The Barong Tagalog is traditionally made with sheer textiles woven from piña or abaca; although in modern times, cheaper materials like silk, ramie, or polyester are often used. It is also known as *camisa fuera*, an outer shirt in Spanish. It fit Sol very well. He looked handsome. I could see the gleam in my parents' eyes; although my mother was already in tears and said she would miss me when Sol took me away with him to the United States of America.

It was a grand wedding, and my family were all committed to making our wedding celebration the best our town had ever known. My oldest brother Albert was the chef, and he planned and prepared all the food. I had total confidence that Albert had everything under control, which made it easier on mom and dad, removing one item from their list of wedding worries. The food and wine were served with elegance, yet in a comfortable setting. Albert took charge of roasting a pig and oversaw the preparation of delicious Filipino and American dishes served to all our guests who were more than eager to partake.

While our family and guests were eating, Sol and I got ready to perform the Money Dance as part of the traditional wedding reception entertainment. This tradition is a fun one, and many cultures choose to include it during the reception. As the newly married couple begins the dance, guests line up to pin peso bills to the bride's dress and to the groom's Barong Tagalog. As we danced, my family pinned a lot of peso bills to Sol's Barong Tagalog to welcome him into the family. And then our *ninong* (godfather) and *ninang* (godmother) pinned money to my dress and to Sol's clothes—to shower fortune on us as newlyweds. Then we opened the dance hall for all the guests to dance.

Time flies when you're having fun, and it was dusk when the last guests finally left. We spent our honeymoon in Baguio City and Banaue, where we saw the exhilarating and historic Banaue Rice Terraces. The Banaue Rice Terraces are commonly referred to as the Eighth Wonder of the World. They are 2,000-year-old terraces that were carved into the mountain of Ifugao in the Philippines by ancestors of the region's indigenous people. They were built with minimal equipment, largely by hand. The terraces are located approximately 1,500 meters (5,000 feet) above sea level. They were fed by an ancient irrigation system from the rainforests above the terraces. To this day, local people still plant rice and vegetables on the terraces, although more and more young Ifugaos do not find farming appealing, often choosing employment in the more lucrative tourism industry that is generated by the terraces. This has resulted in the gradual erosion of the iconic steps, which require constant reconstruction and care. In March 2010, a severe drought made the situation even worse, when vegetation on the terraces dried up completely.

Sol and I stayed in one of the Nipa Huts in Banaue. I believe we paid less than $300 for a week's stay there. We

Part I: Destination: Mosul

also toured Makati and Quiapo, Manila. Sol told me he enjoyed the Jeepney rides in Manila, although the drivers drove *close to a dime*. Jeepneys are jeeps or buses and are the most popular means of public transportation throughout the Philippines. Some Jeepneys have a capacity of 15 people and others have more. Over the years, Jeepneys have become a symbol of culture and art in the Philippines. He told me that he was amazed how the people do this every day without an accident. I told him that I was amazed as well. After our honeymoon, we went back to the province and Sol taught me how to drive a Jeep Wrangler that we bought from my cousin. It was red and it was my first Jeep. I was 31 years old at that time, and Sol was in his 50s. Sol was daring. One day, I took Sol for a joy ride and toured the neighborhood. On our way back home, Sol did not lurch over to grab the steering wheel when he saw an oncoming truck in my lane. He was curious what I was going to do. Well, I changed lanes and got back into the right lane, of course, to prevent an accident, and Sol let out a big sigh of relief. I looked at him and he looked at me and we laughed together. Sol was so much fun, and that's what my family thought, too. He said I was bold, daring, and calm under stress. I remember his words to this day.

In 1984, Sol brought me to the United States of America. We settled in Honolulu, Hawaii, his hometown. It was located about 20 minutes away from downtown Waikiki, where I often walked barefoot. I found Honolulu reminiscent of the Philippines: warm people, green vegetation, fragrant flowers, pineapples, mangoes, palm and coconut trees, mountains, beaches, and—most of all—similar cuisine. Sol took me to his family's vacation home in Makaha Beach. It was a nice house. Sol was not surprised by my quick acclimatization to Hawaii. After all, as I already have mentioned, for me, Hawaii and the Philippines have much in common. Most importantly, they are both islands full of loving and caring people.

Sol took me around Oahu, and we stopped at the pineapple fields and then the Polynesian Cultural Center. The Center is Oahu's top attraction—it wowed me. I have never seen anything like it before. The Center brings to life the spirit of Polynesia through its six Polynesian villages, luau, and evening show. Very impressive. I felt like dancing with them. Sol laughed at the idea, and he encouraged me that maybe someday I could join them. I told Sol that my brother and I used to attend fiestas and always came away as the winners of the dance contests. My brother-in-law taught me and my brother how to dance the Tango, Salsa, Cha Cha, and many other dances.

After a short while I found it interesting that some people would confuse me for a native Hawaiian. All I needed to do was to learn to speak like a local and learn to like eating poi, a Polynesian staple made from the underground plant stem or corm of the taro plant. And I did; I ate poi like a native and started speaking like a local. I got along well with my in-laws. I found common ground with them, especially when I made Lumpia Shanghai or egg rolls and some other Filipino dishes. I felt like a true Hawaiian in Sol's family. I dressed like one and I acted like one. I began to blend in really well. After a month or so, Sol had to get back to work in Kwajalein Atoll, home to the Reagan Test Site. I stayed in Honolulu, and my silent freedom took me to the job site.

After Sol returned to the Kwajalein Atoll, I began to look for work in accounting, which was my education background, or in something that involved these new digital machines, known as *computers*. However, instead, I ended up enlisting in the U.S. Army as active duty—and I found that I was up for the challenge. In short, I went to basic and advanced training in South Carolina, and was initially assigned to Fort

Part I: Destination: Mosul 9

Campbell, Kentucky, which would be my future retirement base 23 years later.

Three months after my duty assignment in Fort Campbell, I received an emergency call from Honolulu. Sol had an accident on base. He had a subarachnoid hemorrhage. It is a life-threatening stroke caused by bleeding in the space surrounding the brain. According to his physician, it could be caused by a ruptured aneurysm, arteriovenous malformation, or head injury. One-third of patients will survive with good recovery; one-third will survive with a disability; and one-third will die. Sol's case fell into the second one: he survived with a disability—but died later on. I flew on emergency leave as soon as I got the approval from the first sergeant and the commander. I went straight to the hospital as soon as my plane landed at Honolulu International Airport (now known as Inouye International Airport, named after Daniel Ken Inouye, who served as a U.S. Senator for Hawaii from 1963 until his death in 2012).

I found Sol lying in a hospital bed and speaking normally. I was relieved when he talked to me and said that there was nothing to worry about. I kissed him on the forehead, thankful that he was alive. However, the following day when I returned to the hospital, Sol showed no sign of remembering who I was. Also, the left side of his body was paralyzed. I was devastated after my conversation with Sol. Just the day before, he kept mentioning my name and said he was happy to see me and how much he loved me. We talked about his dreams for me and our dreams together. He said that I could do anything, be anything I wanted to be. America is the land of the free and the home of the brave, and he said he saw that in me. He said that I could overcome any obstacle that came my way, because I believed in Him, our Creator. I was amazed with his memory and potential recovery. We would be together again and would travel the whole world. But not that day, and not with that Sol. He didn't remember my name or who I was. He kept asking me who I was, and it was difficult to comprehend what he was saying. His speech was slurred, and his words were garbled and hard to understand.

When Sol was released from the hospital, I took leave from the Army to take care of him. We stayed in the family housing at Fort Shafter, Honolulu while I was serving in the Regular Army. Having a great chain of command was the best thing that ever happened to me. How many immigrants like me believed that most Americans have a big heart? I always had. I also believed in my Creator, and I always believed in Angels. My chain of command was made up of my Angels, supporting me—and my Sol—one hundred percent. I remember receiving from my chain of command a Thanksgiving basket of turkey, ham, sweet potatoes, canned goods, nuts, and all kinds of good food. I said I didn't need it, but my chain of command said that they cared for their soldiers and their families. This gesture taught me to give back to the Wounded Warrior Project, because my superiors showed me some love.

Sol never recuperated and passed away in 1986. As a U.S. Army Veteran, he was buried in Honolulu National Cemetery. I visited his grave often because it was close to my base. I brought flowers and cleaned his gravestone every weekend. I could feel his soul watching over me and taking care of me. I believed Sol was in Heaven because I could feel it. I would pray and ask God to take care of Sol and give me the strength and energy to continue through life without him.

In 2003, I remembered my father's words: "Always believe in yourself. Dream big and make your dreams come true!" I did. But there is something I added to my belief. I believe that believing is freedom. Believing in my Creator, in His words, and everything about Him is the best thing that ever happened in my freedom.

Part I: Destination: Mosul

I deployed to Iraq in 2003 in support of the Operation Enduring Freedom (OEF) and Operation Iraqi Freedom (OIF). I have never been closer to God as when my unit was deployed in the town of Ur in southern Iraq. How many of you believe in Karma? I felt like I was reincarnated, that I had been born in that part of the country and had returned. Everything seemed familiar: their food, their culture, and even their language. It was strange but it felt like home to me. If it were my home, would I see siblings from this life? How long had it been since I had been home?

THE HOUSE OF ABRAHAM

Connecting gates at Abraham's house, Ur, Iraq.

I embraced the experience with all my heart and soul while stationed in Ur. I was able to see many things that are mentioned in the Bible, including Abraham's home. The reinforced home had about 32 rooms and was built before Abraham left for Canaan, according to the Bible. I also saw the ruins of Ur, including the ziggurat. The ziggurat is a site of biblical and historical significance. The tour guide who lived his entire life next to the Ziggurat of Ur stated that the step pyramid was built by ancient Sumerians over 4,000 years ago under the rule of King Ur-Nammu. It served as a temple to the Sumerian moon god Nanna. Ziggurats usually contained seven stories: the bottom story would be the largest, the next story would be slightly smaller, and the next would be slightly smaller than the one before, and so on. The construction created the effect of a stepped pyramid. The Temple of Ishtar built by King Nebuchadnezzar II of Babylon was actually built over the ruins of an earlier ziggurat believed by some to have been the Tower of Babel.

THE ROYAL TOMB

I also toured what was reputed to be a royal tomb. The Royal Tombs of Ur is a 4,800-year-old Sumerian burial site of about 2,000 graves located in this ancient city in southern Mesopotamia, which is in southern modern-day Iraq.[1] Sixteen of the graves were designated as "royal" due to the spectacular treasures inside, including gold beads, bronze relics, cylinder seals, musical instruments, and ceramics, as well as artifacts associated with mass ritual. The cemetery was excavated by the British archaeologist Leonard Woolley in the 1920s and 1930s, which sadly resulted in many of the precious

---

[1] "Where Sumerian Rulers Lie: The Royal Tombs of Ur," last modified January 1, 2017, https://www.ancient-origins.net/ancient-places-asia/where-sumerian-rulers-lie-royal-tombs-ur-007294.

The Royal Tomb, city of Ur, Iraq.

relics ending up in the British Museum in London, instead of remaining in their homeland. Only a small number of artifacts from the cemetery can be found in the Iraq National Museum in Baghdad, while the rest are in the University of Pennsylvania Museum of Archaeology and Anthropology in Philadelphia. The City of Ur is historic and deep in my heart. I know I have been here before; I had the exact, same feeling when I reached Baghdad and Mosul. I did not tell anyone about how I felt in this place and chose to keep it to myself. Maybe I was from this world before. Was I a princess? I thank my silent freedom for thinking I was a princess in this part of the world, thousands of years ago. I was free to think what

I wanted to think, and there was nothing else I could think at the time but that of being a princess with a genie who can give me three wonderful wishes.

I did not have a genie, but when I became a U.S. citizen, I memorized and carved into my heart the Preamble to the U.S. Constitution. It states, "We the People of the United States, in order to form a more perfect union, establish justice, insure domestic tranquility, provide for the common defense, promote the general welfare, and secure the blessings of liberty to ourselves and our posterity, do ordain and establish this Constitution for the United States of America." Scholars have stated that the Preamble indicates the Founding Fathers' intention to build a federal government dedicated to protecting the people and ensuring that they always live in a safe, peaceful, healthy, well-defended, and free nation. After I joined the U.S. Army, I felt it my duty to protect my fellow citizens from harm so they can live freely in a democratic country, so help me God.

Fourth of July

Thirteen years later, I appreciate life more than ever after OEF/OIF. It is good to be free, not worrying about enemies lurking right outside the fence. I have been watching the TV every Fourth of July—what a perfect day to reminisce my life's journey in combat as I sit on my balcony and watch the colorful, wonderful fireworks. I have observed that every year Mother Nature blesses the United States with water from Heaven. It is a real honor to welcome rain every Fourth of July, celebrated on the West Lawn of the U.S. Capitol. Last year it rained lightly, and the year before was the same, as it was the years before I moved to DC. I don't recall a year it didn't rain on the Fourth of July. I truly believe that we receive God's blessings more on this occasion, in celebration of our freedom and country—the land of the free and home of the brave.

Part I: Destination: Mosul

It is July 4th, 2016: I see countless people wearing red, white, and blue disposable ponchos. I watch the celebration of the Fourth as usual, like I have in previous years, on TV. I have been living in metro DC for 10 years now and I have yet to watch the Fourth on site. My favorite parts of the celebration include the singing of the National Anthem, the entertainment, the band, and the smiling faces of people who are free. They are free from oppression and sing to the beat without restraint, with the sovereignty and liberty to exercise one's right and the power to sing freely. Basically, I enjoy all the onstage performances celebrating the Fourth. The celebrity host this year is the Emmy Award-winning television personality Tom Bergeron. Tom has hosted the Fourth 2012–2014 and again in 2016.

I like the aerial view of the national parks shown today. I feel proud to see Yellowstone National Park in Wyoming, Montana. I also like seeing Yosemite National Park in the California Sierra Nevada mountains; Zion, Arches, and Bryce Canyon National Parks in Utah; Sequoia National Park in California; Glacier National Park in Montana; Arizona's Grand Canyon National Park; the Everglades National Park in Florida; Mount Rushmore National Memorial on the majestically beautiful Black Hills of South Dakota, as well as other national parks that are very familiar to many Americans. The fireworks have finally begun, and I see red, white, and blue firecrackers in balls that fly and explode in the beautiful skies over the U.S. Capitol.

The Fourth of July fireworks this year are barely visible through the cloud lining as seen from the Truman Balcony of the White House. From my viewpoint, I see fireworks, but not those coming from the Capitol Hill area. The fireworks I see are coming from the Pavilion, where many children and their parents are enjoying the sparkle and crack of skyrockets. Twitter users call out PBS for using old footage of a U.S. Capitol fireworks show. As I said, it seems to rain every Fourth of July, and earlier today, clouds cause poor visibility for those watching the fireworks show in DC, which resulted in some confusion for viewers at home. I was convinced that the fireworks were live until I noticed the sky was clear. I live in northern Virginia, about 35-minutes travel by car from the district.

I work in the district and have seen the Freedom Bell at Union Station in Washington, D.C. It is a replica of Philadelphia's famous Liberty Bell. President Obama delivered his speech in front of the Freedom Bell during the "Let Freedom Ring" ceremony on August 28, 2013 to commemorate Dr. Martin Luther King, Jr.'s "I Have a Dream" speech and the March on Washington for Jobs and Freedom.

It is July 4, 2020: This year's celebration allows me to celebrate my silent freedom. It drizzled earlier in the day but now it is perfect. The Fourth of 2020 was celebrated in a different setting. On July 3, President Trump and First Lady Melania Trump celebrated Independence Day at South Dakota's Mount Rushmore Fireworks Celebrations, the first time in over a decade. It was magnificent. There could have been a larger crowd, but due to the pandemic, Americans have been more careful and are practicing increased safety measures. As I followed the news on TV, the news anchor states there are about 7,000 attendees at the celebration. Due to the COVID-19 pandemic, which has killed thousands of Americans and even more individuals worldwide, American activities have been limited and the wearing of masks and social distancing are being mandated. The following day is the real Fourth of July. President Trump and First Lady Melania Trump are at the White House to host the Independence Day celebration. Again, it is another celebration of Independence Day to be enjoyed from my living room. The weather is truly perfect. There was just a light drizzle, no hard rain. I am never prouder than when watching the U.S. Army Paratroopers team fly, unfurl the American flag, and float down to the ground in a tribute to our country. I have never been prouder of our brave men and women in the military. I am proud to be an American Veteran. The planes and helicopters had perfect weather to fly—and to top it all, it was a perfectly clear night for the magnificent fireworks display. I thoroughly enjoyed all the elements of the show: from when the paratroopers reach the ground to the planes and helicopters flying over DC to the military bands playing and the magnificent fireworks display. I love to hear the narrators introduce the planes as they fly overhead: the planes used during the Vietnam War, the plane that dropped the bombs on Hiroshima, the planes used in the Persian Gulf War, the OEF/OIF planes and choppers, and many others. The last planes to fly are the magnificent Blue Angels and Thunderbirds. They are spectacular and always have been. The pilots flying those planes are my everlasting heroes. I salute you. The protesters from Black Lives Matter (BLM) are also in the area exercising their freedom of speech. Democracy at work!

The foundation of my democracy is my silent freedom. I remember what my father told me: "Believe in yourself; dream and follow those dreams. And believe in the freedom to make those dreams a reality." My mother believed in what my father said, to believe in yourself and to help others as best you can. My mother was spiritually inclined and believed in Lord Jesus's number one commandment: *Love your neighbors as you love yourself.* My mother told me a person should pursue their own goals, which would fulfill yourself as a person. In other words, you can think what you want to think. I am free to dream to the fullest extent of my imagination. If I choose to become a physician, an astronaut, a lawyer, a business engineer, an accountant, or a computer hardware and software designer, I am free to do so; and I will be willing to do what it takes to fulfill that dream. If I am a filmmaker, then I am free and should be willing to create a story as far as my imagination can take me and deliver the best film ever. If I want to become the best graphic designer, then I can maximize my dream because I have freedom that will allow me to do so. If I were a waitress, a child provider, then treating the customers and parents like I want to be treated plays perfectly. Whatever I want to become is possible. Silent freedom is where great changes begin.

In my spare time, since I came back from the war, I watch TV shopping channels online, such as HSN, JTV, and QVC. These were the channels that I knew of when I came home. HSN and QVC have been around for years, and I am amazed at the number of shoppers who have been following them now. On these channels, there is the liberty to sell just about anything, such as kitchen wares, electronic items, garden tools, exercise equipment—not to mention items from the current trends in fashion and jewelry. The liberty to sell things on TV allows shoppers to open a credit card account with the company and charge their purchases with just a flick of a finger. The shoppers are free to explore and choose what they want to buy—the freedom to choose and the freedom to enjoy life. America is the land of the free and the home of the brave, as my father kept repeating while he was alive.

I also like to watch shows that take me to places I have never been to before. For example, I was watching a tour of Palestine and saw a particular site with paintings on the wall. It reminded me of my travels through the Baghdad International Airport. I was awed by art on display there, and the artist's imagination. I cannot believe they referred to that part of Baghdad as the Yellow Zone and prohibited most U.S. soldiers from seeing those exhilarating paintings on the wall; it was a canvas for artists to paint and express Iraq's natural beauty. I have never seen anything like it before. Whoever painted the wall is truly gifted. It was a marvelous painting! The paintings on the Palestine wall are quite different, mostly depicting the former Palestinian leader Yasser Arafat. Nevertheless, I enjoyed the adventure on TV led by an American tour guide.

The shopping malls are also a huge feast for my eyes. As I browse the stores, I see nice clothes, fashionable jewelry, shoes, and cosmetics. The evening gowns are spectacular, and high-heeled shoes that match these sequined gowns are stunning. There have been several times that I imagined I was Cinderella wearing a beautiful gown with glass slippers and attending a wonderful ball, only to run away just before midnight, before my dress turns back into rags again. I am free to dream; I am free to explore. While I was deployed in Iraq, when shopping on the Internet became common, soldiers purchased most items online. In fact, shopping online was probably the number-one method of shopping because there was nowhere else to shop, except for the little bazaars. I noticed that most troops shopped online for Tylenol, Aleve, shirts, or anything else that they could not find in the exchange or bazaars.

Product choices were very limited in combat theater when I was deployed. For example, when I looked for a long-sleeve shirt, my only color choices were black or blue. Either there were two to choose from or there was just one available. That is why it was shocking for me to see so many choices when I repatriated to the U.S. I was amazed to see almost everything in one place. There is a lot of freedom of choice. Every time I wanted to shop for a pair of shoes, there were other colors available. Most of the time, I end up buying the same shoes in all the available colors. I have been a different person since I came back to the U.S. after the deployments. My perspective in life has changed; my actual life has totally changed. I used to buy a single pair of shoes, which I would wear for the whole year. Now, I have multiple pairs of shoes, in all kinds and all colors. Freedom in entrepreneurship has allowed individuals to further their discoveries and help others improve themselves and their undertakings. Mark Zuckerberg's co-creation of Facebook, for example, has caused users to spend an average of 10 hours or more each month on the social networking site (including some of my relatives). My son, who is a millennial, and his friends spend about 15 hours connecting with their friends and families about their adventures, their new discoveries in filmmaking, and what they see going on around the world.

We are free to choose. I vividly remember the words of Christ in the Bible, according to Mark 8:34: "And when he had called the people [unto him] with his disciples also, he said unto them, Whosoever will come after me, let him deny himself, and take up his cross, and follow me." Jesus is not mandating us to deny ourselves, take up the cross, and follow him. Rather, "Whosoever will come after me… will have eternal life." These words have been my mantra, the real reason for my existence on earth: my hope for eternal life. I chose to follow Him, my Creator, my Refuge, my Savior.

As if to confirm my silent freedom to follow Christ, on the eve of Labor Day, September 4, 2016, Father Zacarias preached in Our Lady of Angels Mass that carrying the cross means to sacrifice. He preached, "Yes, we are free to choose. If we choose to follow God, then God will be with us. He will be our constant companion and will be with us and will not leave us to do it alone." I like his homily. It reinforces my faith to follow Jesus. I remember someone once asked me why I go to church a lot. My following of Christ began in my early childhood when my mother took me with her every Sunday morning to profess our faith. Going to church on Sundays has been a part of my life, not because I was brought up that way but because I chose to follow Jesus. There had been so many times in my life when my faith was challenged and there had been several times when I failed these tests, but I would always find my way back to Christ.

THE WAR ZONE

My silent freedom to serve my adopted country and protect the people took me to the war zone after President George W. Bush declared war against Iraq on March 20, 2003 at 0800 Zulu Time (also known as military time). I deployed with the 101st Airborne Division, Air Assault (AASLT). On day one of the war, some of my fellow soldiers and I were caught in Arifjan, Kuwait doing personnel service support to the transportation battalion. It was set for three days but was shortened to two days; it was the first really bloody day in the war zone, day one of OIF. Staff Sergeant (SSG) Hayes from our team did his early morning run on that day at the airport to pick up a new incoming soldier for his unit. The wait was out of normal range. After all, it was war. It took three hours or so to dispatch a vehicle. But I was willing to wait longer than three hours, provided we were able to get back to Camp Udairi, Kuwait. My unit was tentatively based in Camp Udairi and we were trying to get back there on the first day of the war, OEF, which became OIF. Mission: to liberate Iraq from the authoritarian rule of Saddam Hussein and destroy his weapons of mass destruction (WMD).

We were now on our way back to Camp Udairi. About 20 miles away from Doha, Kuwait, our Humvee made a hissing noise and the hood started smoking. SSG Hayes stopped the vehicle and discovered that the hose was displaced and would disable the Humvee. Deep in my mind I couldn't believe that such an incident had happened. First, the chaplain was with us, a man considered very close to God. With his faith and our faith, everything should go perfectly according to plan—right? Second, our Humvee was perfectly fine when we arrived in Arifjan, and now there were mechanical issues. We had been performing preventative maintenance to ensure that it ran smoothly. So why would it be stopping now? Was it a coincidence or Murphy's Law? These were trying moments. We broke down in the middle of a bridge, of all places, on the first day of the war. We had to act in defensive position, taking cover between the wall of the bridge and the Humvee. Sergeant (SGT) McKinney and Specialist (SPC) Davis, Jr., were facing east; some soldiers were facing west. I was right beside SSG Hayes trying to fix the displaced hose with a piece of a meal, ready-to-eat (MRE) wrap and his left bootlace. I positioned myself east. The chaplain sat between the wall of the bridge and the other Humvee. I asked him with a hint of a joke, "Chaplain, do you have a special prayer for an occasion like this?" He looked at me like he wanted to say, "Have faith, my child." But he didn't. I went back to SSG Hayes and looked at his progress in reconnecting the hose. In fact, he was making progress. The chaplain was reading a book. We used a cell phone and called the transportation battalion in Arifjan for a wrecker. We had an affirmative answer. We gave them our location, not sure of the exact grid coordinates. An eight-digit coordinate could have been a sweet mercy. About 45 minutes later, we called to follow up. The entire battalion was on Scud missile alert. SSG Hayes was successful with the hose and we drove down a little farther, about 15 more miles. We knew it was a temporary fix. SSG Hayes said the MRE wrap would not last long. We pulled aside at Highway 80. Someone found a piece of rubber that seemed to offer a perfect fix to our situation and might allow us to get back to Camp Doha. We planned to wait for the wrecker or the rescue team over there.

Part I: Destination: Mosul

I decided to call my commander on a cell phone to give him my status report. He sounded anxious to hear from me. He always gets excited when he hears from one of his soldiers, especially on day one of the war, and out of the area of operation (AO). I told him that we had a vehicle break down, and we called a wrecker to rescue us. I asked him the status in Camp Udairi, and he said they were in gas masks due to Scud missile alert. I was just glad he answered his phone. Sometimes it was hard to get through with a cell phone. As our call ended, he told me to take care of myself and do anything I can to get back. That sounds like my commander. I said, "Roger, sir, hooah!" *Hooah* in the Army means, "Yes, no, I agree with you, I disagree with you, or, whatever." I agreed with my commander. I couldn't wait to get back to Udairi and be with my unit. I looked at SSG Hayes and told him about the Scud missile alert. Once again, he was more focused on the hose. SSG Hayes succeeded in fixing the hose by wrapping a piece of rubber and tying it with a string that we found on the side of the road. We hoped and prayed that we would make it back to Camp Doha, about 36 miles from Highway 80. I glanced at the chaplain. He was very relaxed. I always have faith in chaplains. I believe that they could almost make miracles happen.

We reached Camp Doha with beads of sweat. We felt that we pushed the Humvee with all our might, with SSG Hayes' might. He kept telling us that he was a legend in my unit. Truly he was. I told him he was my hero, and he was. He saved us by taking us back safe and sound to Camp Doha, where we waited for rescue. The gate guard at Camp Doha did not let us go through. We were told that a Scud missile landed somewhere in Doha. It turned out later that it was a dud Scud missile. We were directed to park in the sand parking lot and wait for the wrecker. It was a tiring day. SSG Hayes took a nap on his belly, on the Humvee's hood.

He was tired, having driven to the airport at 2100 the other night, waiting for several hours, spending the night there, and then driving back to the unit with a new, incoming soldier. I looked at the chaplain who finally said that he's not coming with us to Camp Udairi. I don't blame him. He had enough for one day. I took a nap in the back seat of the Humvee. Then I overheard the guys talking to NBC reporters, Rick and Allen. Great, I said. Then we heard the Scud missile alert, and everybody started putting on their gas masks, including the civilian reporters. We donned our Jayless suit, the modern nuclear, biological, chemical (NBC) suit. I decided to take my desert camouflage uniform (DCU) trousers off in order to be less hot. I had time to do so. I told SSG Hayes that it would be better if we remained in the Jayless suit for the rest of the evening. He agreed that it was a good idea. The NBC reporters kept their camera rolling while we were donning our Jayless suits. Both of them were also in gas masks. We had been wearing gas masks for about an hour when we heard the "All Clear" coming over the microphone at Camp Doha. Another bad hair day after a Scud missile alert.

It was almost 2100 hours when the rescue came. We finally made Plan A and B scenarios. The chaplain did not come with us to Camp Udairi; he decided to stay in Camp Doha. SSG Hayes drove us back to Camp Udairi with the rest of his S-1 (personnel administrative support) crew. It was really touching to see all of his crew show up to rescue him at a time like that. What a team. We reached Camp Udairi around midnight, after getting lost several times. We found our way after passing through Camp New York, then Camp New Jersey. The leadership's idea made sense to me to name the camps after the U.S. states. It's easier to remember.

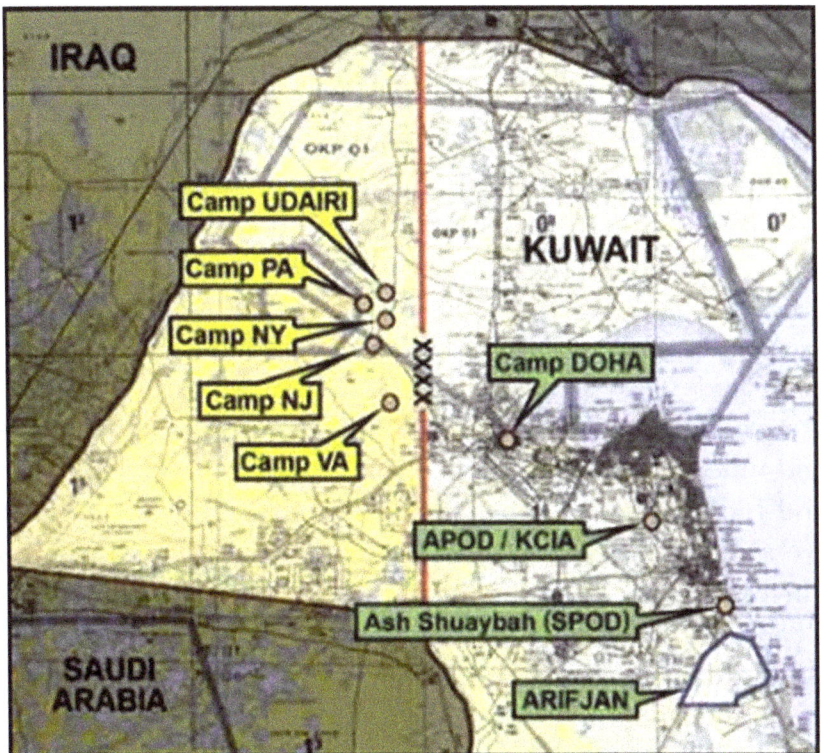

The map delineates the locations of U.S. camps in Kuwait before going to the battlefield in Iraq, OIF 2003.

Upon reaching the gate to Camp Udairi in the middle of the night, we drove to my unit and found it in a dimly lit, sandy section of the camp. We were fortunate that there wasn't a sandstorm that night. The chaplain must have prayed for us. During a blinding sandstorm, there is zero visibility. Regardless of distance, I would recommend staying in the nearest building until the storm is over. As I glanced back, we found out that our battle buddies in the other Humvee were not trailing us. *Battle buddies* are soldiers assigned to us, and each of us is expected to take care of each other in and out of combat. It is expected that we should not leave our battle buddies behind. We got worried, and so the legendary

SSG Hayes planned to look for them around the place. I suggested that he go back to the gate and start from there. After an hour, SSG Hayes showed up with the lost Humvee, full of very energetic soldiers; one was mine, SPC Davis, Jr. We found out later that SGT McKinney and SPC Davis, Jr., had to leave and join their casualty liaison teams that supported our brigades in the 101st Airborne Division (AASLT). I was scheduled for a mission the next day to provide promotion evaluation and personnel service support for the Bastogne 1st Brigade.

At about 0200 hours, we heard a Scud missile alert. We had nine seconds to don our masks, snatch our Jayless suits, and hit the bunker, which was located in front of the tent. We always had our weapons and ammunitions with us—it's our best buddy ever in combat. Someone finally called, "It's clear!" and we got out of the bunker. What a nightmare and what a day. Yes, we had had our share of that day too, just like the chaplain.

In April 2003, my unit commander sent me an email that referred to a master sergeant (MSG) selection board release. He was not clear. He said he saw the list and that congratulations were in order. I did not take it seriously and I responded without mentioning the board release. On April 27, 2003, the Department of the Army (DA) officially released the board results. I was selected for promotion. All of a sudden, I saw a light at the end of the tunnel. It was a new life, with new hope, new freedom. I was so happy. My selection for promotion suddenly brightened up my life and gave me new purpose. I was still recuperating from the sad news that my husband was seeing someone else. I remarried in 1986 to my husband, a fellow soldier I met in Honolulu, Hawaii after Sol passed away. He was also assigned to Fort Shafter with the Corps of Engineers. His job took him all over. One time, his unit went to Samoa for a humanitarian mission, and he got a commendation award for it. He performed heroic deeds and was such a good man, otherwise I would not have married him. It's the reason why I was caught off guard and traumatized when he cheated on me, which led to our divorce. I cried every night and lost my appetite. I lost 20 pounds and I noticed hair loss. My supervisor, who was also a warrant officer, has been a great mentor since day one of my husband's betrayal. One day, I reported to his office and told him that I was done crying and ready to move forward. My supervisor said that it was about time I wake up. I did not know that he had been observing me and counting down to the time when I would begin to accept the devastating betrayal. My promotion marked a turning point in my life, one of hope.

I will never forget the phone call from husband when he told me he was seeing someone else. This never occurred before and I was baffled. After 20 years of happy marriage, without fights, and full of love, trust, and respect, my husband had betrayed me. It hurt. He confessed to me that he met her online when he was looking for someone new. This woman was also married at the time. They started dating after I was reassigned from Fort McCoy, Wisconsin to Fort Campbell, Kentucky. I asked him on the phone what he wanted to do now, and he said he did not know what to do, and that he would tell me all about it the next time he dropped off my son for his summer vacation. It felt like a bomb hit me when he said that. I tried to forget him and his betrayal while I was deployed, but the good memories kept on coming back.

On June 10, 2003, I decided to join a group for a Morale, Welfare, and Recreation (MWR) tour to the ancient town of Nineveh, where we visited the tomb of Jonah. My unit provided us with a break occasionally when it was safe to do so, and we got security briefings a thousand times before departure. It made a lot of sense. After all, we were still at war. Downtown Mosul reminded me of a certain part of Manila, Philippines. I felt very confident but had to remind myself that I was at war in Iraq and was a part of Operation Iraqi Freedom. Downtown was busy with people going to work, kids walking to school, and all kinds of people out and walking around.

We climbed the steps of the acclaimed tomb of Jonah. His tomb is perched on a high mound, containing many layers of history. It was an ancient Assyrian temple and palace, a site of devotion for the Jewish people, a Christian church, and a 12th century mosque. According to history, in 1924, a grand minaret was added by a Turkish architect who described the glow of the site as God's gift to Mosul. It was renovated in the 1990s under Saddam Hussein and was a popular destination for pilgrims. We were informed that the place was converted to a mosque and one could not enter unless they were Muslim. I did not get to enter, but I saw and was able to take pictures of the place from the outside. Five- and six-year-old Iraqi children surrounded us, curious about what we were doing there. Some of these kids were trading dinars for dollars. I asked myself, how did these kids have money? Were they not just manipulated by their parents? Maybe. Outside the temple, there were three sidewalk vendors selling sodas, candies, cookies, and souvenirs. Again, it reminded me of a typical Asian country, such as mine, especially in the provinces where children have to work to help support their family. When we returned to the base, I couldn't thank the Lord enough for the opportunity to serve my adopted country. I never felt closer to Him than at that moment.

Although I too have had misfortunes, they are nothing compared to Jesus carrying the cross. Twenty years ago, my husband and I met in Honolulu, Hawaii. Like every couple, there was a memorable place to meet that special someone. The attraction started from the commissary in Fort Shafter,

Hawaii. I was holding a tea box and he greeted me with "How are you doing?" I looked up and responded sweetly that I was doing just fine, thank you. That was it; I paid for my groceries and headed to the Post Exchange (PX).

I was just entering the PX when I saw the same guy leaving the PX. At that instant, I felt attracted to his physical features and I told myself that I will marry this man. Time went by and we started dating. As I predicted, and true to my word, I did marry him. We were married before a judge in Honolulu, Hawaii, followed by a quiet wedding ceremony in Fort de Russy, Honolulu, Hawaii. The ceremony was conducted by Reverend Father Phillip Floersh, the assigned chaplain in Fort Shafter and Fort de Russy. We celebrated our wedding at the club in Fort Shafter, Hawaii. It was attended by about 50 individuals, who were our friends and co-workers. Rebecca and Mike were our bridesmaid and groom's sidekick, respectively. We had four pairs of principal sponsors. We call them *ninong* and *ninang* in my country. We honeymooned at Turtle Bay Hilton in Honolulu, Hawaii. We also had fun with our friends Rebecca and Mike at Makaha Beach, on the northern side of Honolulu. In 1988, my husband left Hawaii one year ahead of me to attend the Officer Basic Course in Fort Rucker, Alabama. Then he proceeded to his Warrant Officer Course. I was the first one to salute him after graduation, so he gave me his silver dollar coin. It has been a tradition in the Army for a fresh officer graduate to give a silver dollar coin to the first enlisted person to salute the graduate.

His first assignment orders as a warrant officer said that we would be going to Fort Belvoir, Virginia. That was sort of a dream come true. We were preparing for it and were so excited to go to Fort Belvoir, Virginia, only to find out about a last-minute change. We did not go to Fort Belvoir. The orders were changed from Virginia to Fort Benning, Georgia, where we would be staying for the next three years, and where I conceived my one and only child, my son Christopher. As warriors, we had to be flexible and keep marching on. We could not decline the orders unless there was an emergent situation that would prevent us from going. But we did not have any excuses. We were blessed with good health and so we followed the military orders.

We had two dogs named Cinnamon and Chichi in our new home. Cinnamon was part Shelty and Chichi was a Chihuahua that I found at the post's kennel. Chichi was housebroken. She did spit blood when I found her. I paid for her veterinary care inside the military base and took her home. I gave her a bath and cleaned her really well. We liked Chichi a lot. In 1991, my husband received an order and was called to serve in the Persian Gulf war. I wanted to go with him of course; after all, we were both warriors. But he deployed and I was left alone, with Chichi the dog. I was worried about my husband in the Gulf, and he claimed that he was worried about me too. The war ended and we brought our troops home safe and sound. It took a week for my husband to adjust after returning from the Gulf War. He would get up at night screaming, "Don your masks and jump into the f***ing bunker!" I had heard of the possibility of this happening to soldiers returning home from the family support group (FSG), so I calmed him down and told him he was okay. He settled down and finally got back to a normal routine about two weeks later. He said he missed me so much, and so did I. Cinnamon the dog came later on, after my husband came back from the Persian Gulf.

One evening we planned a vacation to Manchester Missouri to visit his parents, brothers, and sister. He had two brothers and one sister. We enjoyed our stay in Sam's house in the country. Sam is Jennie's husband, my sister-in-law. Then we went to the cabin on the mountain, enjoyed the swing set, barbeque, and many other activities with the family. We had so much fun that I began to feel a little nauseous and tired, so I decided to take a nap. It must have been from running around too much, I said to myself. It had been two weeks and our vacation was up, so we drove back to our little home sweet home in Georgia, near Fort Benning. I missed my period during this time. I felt concerned, so I went to see a doctor. I was afraid that I had cancer but instead, I was told that I was expecting a baby. My jaw dropped. It was hard to believe. I was almost 40 years old, and the doctor was saying that I was having a baby—a baby! I was so excited and broke the wonderful news to my husband. He was all excited, too. He had two sons from his first marriage, Cliff Jr. and Patrick Henry. They are very smart kids who earn high grades in school. My two stepchildren were living in Puyallup, Washington State at the time. We had to inform our relatives of the new, incoming child—and they were all excited.

A Permanent Change of Station (PCS) order came to inform us that we were going to Fort Hood, Texas in November 1992. I was seven months pregnant at the time. We had to present a note from my physician saying that it would be all right for me to travel. My doctor advised me to stop every two hours during the drive. And we did. My husband and I had to apply for permissive temporary duty (PTDY) to look for a house to stay in, in Texas. We then returned to Georgia and prepared for household pickup and put our house up for sale. We drove two cars from Fort Benning, Georgia to Fort Hood, Texas. It was not too bad. My husband told me I did well considering that I was very pregnant. We found a cute house in Copperas Cove, Texas and decided to rent it. My son was born in Fort Hood, and we reared him in Copperas Cove. Christopher's nickname is CJ, and it stands for Christopher John. Christopher is for being a good traveler and John (the Evangelist) is for Lord Jesus's favorite apostle according to the Bible. I love Saint John, the Evangelist. He was a witness to many of Jesus' miraculous works. My son was about 10 pounds when he was born, the largest baby born that day according to my nurse. I had a Cesarean, and my son's dad said that Christopher peed on the nurse after the umbilical cord was cut and when she was handing him over to his dad. That was the euphoria of the day, and Christopher became the nurse's favorite child during our stay. My husband and I were very happy and proud of our new child. While in the hospital, the nurse on duty put my son in front, being the largest baby. I was thinking that he seemed to be a born leader, protecting the other newborns while we were in the hospital.

After being stationed in Fort Hood, Texas for one year, my husband's unit was disbanded. We received another set of orders to go to Fort Lewis, Washington. Things started to change with my husband. I was not aware that he had had a fling with a fellow soldier in Fort Lewis on his first assignment there, which he later disclosed to me. He was already married to his first wife and had an illicit affair with one of the female soldiers who was from Japan. She was also married. My husband confessed to me that he loved this female soldier and he asked to be transferred when she had gone absent without leave (AWOL) with her husband. I was not aware what was running through my husband's mind. I thought we were happy together, especially with my stepchildren living nearby and able to visit us so conveniently, as then we were all residing in the state of Washington.

One year later, my husband dropped his retirement papers in Fort Lewis, and we had to make plans after his retirement. He stood up proud and tall in front of the soldier formation that honored him and his fellow retirees. It was an unforgettable event, as it signifies the next life after a 20-year-long, illustrious military career for him. I did not think about my military career at the time and the retirement because I was afraid it would demotivate me. I know myself and what motivates and what demotivates me, and one of them is looking at the end of the tunnel when I still have a long way to go before reaching that tunnel. I was in the Army for only 6 years by the time my husband retired, so I had a long road ahead of me before the 20 years were up, or maybe I would serve more than 20 years. God will let me know when my time is up.

Our plan after his retirement was that he was going to teach part-time and go back to school to pursue an electrical engineering degree, in addition to his Bachelor of Science degree in computer science, where he graduated cum laude. I was willing to support him financially, in addition to his retirement pension, while he was in school for the four-year engineering degree. We both knew he could do it. He was smart. But everything fell apart when his friend, a fellow soldier in the Army, gave him a lead for a new employment opportunity in Michigan. It broke my heart to let him go, but at the same time I did not want him to blame me later for not letting him fulfill himself by not taking that job in Michigan. Holding him from such an opportunity did not fit in with my definition of freedom. He took the job, and I was left with my son who was about one year old. My oldest stepson stayed with me for a while. I guess my husband was too immature at the time to realize his responsibility, and it just irritated me. But my oldest stepson turned out to be a very responsible young man. He is now a successful certified public accountant, married to a beautiful and successful real estate professional, and they have one very smart kid, Evan. My younger stepson also finished his mechanical engineering degree and married a wonderful wife. Both live in Texas because, according to Pat, they love and take care of their Veterans in Texas. Pat is a disabled U.S. Army Veteran.

We both served in OIF. He was in Balad, Iraq, and I was in Baghdad. I was unable to attend his promotion to first lieutenant due to a transportation issue, which is common in combat theater. He got out as a captain.

I communicated with my husband over the phone while I was stationed at Fort Lewis and he was working in Michigan. It was not easy to take care of my son as a single mother while actively serving in the U.S. Army, where we go to the field and are pulling duties at the same time. I had a family care plan, of course, but due to the hardship, I invoked my silent freedom and told myself that I will never forgive my husband for this. But when I received my orders for Korea, I had to leave my son with him, so he would enjoy stability and be taken care of in the United States. I sent allowance payments for my son's care, and my husband was there when I came back. He gave me moral support, and I was thankful that my family was still intact.

My next assignment was Fort McCoy, Wisconsin upon return from Korea. This duty assignment allowed me to see my family often by meeting them halfway and staying in a hotel for the weekend. We did this until we bought a beautiful lakeside house in Michigan, which was located near the police station, post office, Eastern Michigan University, and grocery and hardware stores. My husband's hobby is tinkering with used cars, and he was very happy to find a car parts store nearby. That was a blessing for him. Also located nearby were St. John the Baptist Catholic Church, the University of Michigan, St. Joseph Hospital, and the Veterans Medical Center in Ann Arbor, so we felt blessed. I had to go home constantly to decorate our new house. I thought we had everything: a beautiful house, a boat, a jet ski, and—most importantly—a happy family. With my son, my husband, and my stepchildren in the same household, and the fact that we no longer needed to travel to Washington State to see them, plus good in-laws—I thought, we've got it made. Then I thought again, *I've* got it made.

Fort McCoy, Wisconsin was a paradise for me after coming back from Camp Henry, Korea. I got to go home twice a month to see my family and got to live like a family. We ate together, stayed together, saw places together, played together, and did what a happy family does. It was in Wisconsin that I saw the attack on the World Trade Center in New York City. I was working in S-1 on the first level of the building, and the Operations shop was located on the second level. The TV was at the Operations shop, and the operations sergeant called me on September 11. He yelled that the Pentagon was on fire! And then we saw the airplane hit the Twin Towers in New York with the plane diving through the building. We saw people jumping out of the tall building, the fire, and all the chaos that was happening. It was a horrible sight and enough to agitate us. It was a day to remember, September 11, 2001. We watched Aaron Brown as he covered breaking news attacks against New York and Washington, live from CNN NY rooftop. Our eyes were glued on him and the news that he was delivering to the people on that morning. Later, the news reported: On September 11, 2001, 19 militants associated with the Islamic extremist group al-Qaeda hijacked four airliners and carried out suicide attacks against targets in the United States. Two of the planes were flown into the towers of the World Trade Center in New York City, a third plane hit the Pentagon just outside Washington, D.C., and the fourth plane crashed in a field in Pennsylvania. Often referred to as 9/11, the attacks resulted in extensive death and destruction, triggering major U.S. initiatives to combat terrorism and defining the presidency of George W. Bush. Over 3,000 people were killed during the attacks in New York City and Washington, D.C., including more than 400 police officers and firefighters.

The attack defined President George W. Bush's presidency indeed as soldiers would be deployed to Iraq later. I received my PCS orders from Fort McCoy to Brussels, Belgium. I was very excited with the assignment, only to find out that my new job position had changed. I had two new choices: either Germany or Fort Campbell, Kentucky. My husband was quick to select Fort Campbell, Kentucky because we would be able to see each other more frequently than if I had been stationed in Germany. I agreed. I never expected that this assignment would break my marriage. A month after I was assigned in Fort Campbell, he found somebody else through the Internet, and she was still married. I was devastated. It was a challenge of my love, faith, and the way I viewed an individual's freedom to choose.

Sometime later, I received orders to deploy to Iraq. I welcomed my first deployment wholeheartedly with the 101st under OEF and OIF. We were staged in Camp Udairi, Kuwait. We set up our work tent next to the Mayor Cell to connect with their power. Everything was a luxury, from water, Internet connections, to the phones. Even letters in the mail were considered a luxury. We always looked forward to hearing from home through snail mail while we were waiting to deploy to Iraq. I spent Lenten season and Easter Sunday in Camp Udairi, and it was very holy. We did not have a priest, but the Lay Eucharistic Minister (LEM) did an outstanding job. He read a thoughtful verse and had a good understanding of the Holy Gospel. According to him, he did not convert to Catholicism until the Persian Gulf War broke out. He attended a Catholic Mass, and it made a deep impression upon him.

One Sunday morning, about 1100 hours after the Catholic Mass, we heard shots outside. Everyone rushed into a defensive position. It was disappointing, but I did not have ammunition at the time. I had an M16A1 without ammunition.

One soldier who was a specialist gave me a magazine with ammunition and I thought that was kind of him to give me power to defend myself or somebody's life from whomever was causing that fire. The shooting was not very far away. After about 20 minutes, we all calmed down and found out that a foreigner drove a truck into 11 American soldiers who were waiting in line at the PX. The culprit was shot by a specialist from HHC Division, which gave the victim a sucking chest wound. The culprit survived and was air-evacuated to USS Comfort, which supported OEF/OIF. The American soldiers were taken to 86th CSH for treatment. No one was seriously hurt. Some had cuts on the knee and lower leg; others had concussions. Later on, in the month of April, as a Human Resources warrior, I was asked if those patients who were victims of a terrorist attack would qualify for a Purple Heart. According to the Army Regulation (AR 600-8-22), a Purple Heart could be awarded to battle injury casualties and as recommended by the chain of command. We were in a battle during that time.

I remember on April 21, 2003, I left Udairi for Camp Adder. I was in the 86th CSH ground assault convoy (GAC) to accomplish a mission as a Casualty Assistance Liaison for the 101st Airborne Division, Air Assault (AASLT). Destination: LSA Adder in Tallil Air Force Base. The city of Ur is 10 minutes away from Camp Adder, and Camp Adder is 90 miles north from the border of Kuwait. Adder is 10 minutes away from Nasiriyah, which is relatively well-known now in the United States, since the coalition forces rescued PFC Lynch from a small hospital in Nasiriyah, Iraq. I remember Lynch before I saw her on television in Camp Udairi. Someone called the hospital asking for Private Lynch and stated that Lynch was missing and that they had called each medical facility. At that point, not one of these facilities had admitted Lynch. I remember writing Lynch's name in my notepad and looking at the patient list. Indeed, there was no Lynch admitted in the hospital.

The next day, as I was watching Fox News in the MWR tent, I saw Lynch on TV. She was a supply specialist who was seriously injured when her convoy was ambushed by Iraqi forces during the battle of Nasiriyah in OIF. She was reported to have been seriously injured and was captured by the Iraqi forces. There was big medical coverage about her as she was subsequently rescued by the Special Operations Forces from a hospital in Nasiriyah, which was about 10 minutes away from our camp. The name registered with me very well, and I could not believe that Lynch on TV was the same PFC Jessica Lynch they had asked me about the night before. I kept the notepad I had for a long remembrance. I told myself that Lynch should be groomed for Congress and fight for her constituents, the Veterans, and the citizens of the United States. Her tenacity, faith, and courage to stay alive while held as a prisoner of war (POW) would be excellent qualities to fight for what is right and survive the daily battles in our U.S. Congress.

On that evening, Dr. Ali took me for a walk around the Living/Sleep Area (LSA) Adder. It was about a mile walk. I was in my physical training (PT) shirt and shorts. We moved behind a Milvan and sat on the dirty tents and tent poles. At about 1830 hours, he excused himself and said his prayers. He prayed four times a day. He was Muslim. We talked again after he prayed, and he smoked a cigarette. He kept singing the same song while we were walking. I finally asked him to translate the song for me, and he said it was about a woman and man who fell in love with each other and everyone in their families opposed the relationship. It sounded like Romeo and Juliet to me. He asked me how he should tell an American woman that he liked her. I told him just to tell her what was in his heart. I was glad he was not referring to me.

But he continued and persisted that he would like to say the right words to this American woman. He said he tried to tell her, but she just ignored him.

Finally, I asked him if I knew the woman, and he said yes. I asked who? And he laughed like a little boy. He said he knew that was coming. We both laughed. I finally guessed it right. It was the young blonde soldier in the section. That was a relief to me. I did not want any relationship beyond pure friendship at the time. Mohammed came to me every time for anything he wanted to know and asked me to say something to the other girl in my shop. He even picked me up every day for dinner at the DFAC. He knew I was the only one he could talk to back then. Nobody knew him better than I did, with the exception of sexual intimacy. We never had that. Mohammed and I are both from the continent of Asia; he is from the West and I am from the East. Besides, he told me I was like an Egyptian woman, more like a mother to him because I was feisty and displayed those common maternal traits.

We had tea, which they called *chide* in Arabic, one night in Mohammed's tent. Three of us were there: Mohammed, the other girl, and me. He made good, strong chide. After an hour, I had to excuse myself and left them both in the tent. The following day, Mohammed told me what happened when I left them. He told me everything, including his relationships with other women in the hospital. Relationships, which excluded sex. He trusted me. Sex was forbidden during the deployment. I really enjoyed my work at the 86th CSH, especially when we jumped from Udairi to Adder. The nurses and doctors were truly professional and some of them became my friends. Captain (CPT) Quintana and CPT McKoy were in the same sleep tent with me, in the dirt, at Adder. There were nine of us in that tent and we all got along together. I remember MAJ Robinson from ICW2. She was very pleasant to talk to. She was the Chief Ward Nurse and took care of the patients, both in Udairi and in Adder.

Another soldier I will always remember was SPC Wade, who went running to rescue me from the dogs, and SSG Figueroa who rescued me with the Gator. It happened one pleasant morning when I was walking on the side of the airfield and was ready to double time (run). Suddenly, nearly two dozen street dogs ran out of the building ruins and chased me. They were barking loudly and woke everyone up in the tents and the neighboring units. I grabbed rocks and tree limbs to protect myself but to no avail. The dogs made a circle and surrounded me like canine terrorists. It scared me. I never saw anything like it before. Their fangs were big and sharp, and saliva was flowing from their wild fangs. SPC Wade looked in my direction and just stared at me. Later that day, she said that she thought I waved at her, and did not see the limb in my hand. She realized that I was in big trouble when she saw me frozen in the middle of the pack of dogs, so she came running toward me.

SSG Figueroa also saw what happened, took the Gator nearby, and drove the Gator toward me. He told me to hop in and watched the dogs because they might lunge at me. And I walked backward to the Gator while watching the dogs coming in closer and closer. As soon as I sat down, SSG Figueroa started driving away from my potential killers, and I saw that the dogs were still chasing us. They finally disappeared as they approached the bystanders in the tent world. Some of the bystanders were in a ready position to shoot the dogs if they continued to attack us. We were glad that the dogs stopped and ran away. I was relieved as soon as I got off the Gator, and everyone was happy that I got back safely. That incident was my excitement for the day. I thought I would die from dog bites instead of being killed from gunfire because, after all, we were in a war zone.

I could visualize the headlines: *Soldier Killed By Iraqi Dogs, Not From The Gunfight.* I would never forget how the dogs surrounded me. Those dogs could be police dogs if properly trained. I could have brought all of them back home and given them to the military police (MP). I called my commander and first sergeant (1SG) who were already in Mosul. They were petrified and told me to be careful. The news spread like wildfire in my unit and might have been part of the daily battle update brief.

Our tent world was not too shabby. Some sleep tents were decorated nicely, especially in Major Summers' tent. They put two artificial yellow and white roses on the tent pegs outside the front door. There was a sign that read "Sleep tent of the month" posted on the front door. I thought it was cute. The mailroom was attached to the old hospital. Mail run was from 1300 hours through 1500 hours. We always like mail calls. At least there was something to read in addition to emails. Additionally, we received care packages from our families and friends in the United States. Some soldiers corresponded with U.S. school children. The children sent banners, flyers, letters, and care packages to thank the soldiers for being heroes. One girl asked how I liked my job. The others asked what it is like to be in Iraq fighting the war, and others asked how many enemies I have killed. The children's curiosity was what I needed to keep my faith and morale high. I believe that young children are the future leaders of the United States and, therefore, should be kept abreast of patriotic deeds that adult Americans have been doing, including OEF and OIF. On the verge of tearing down the hospital to retrograde back to Fort Campbell, soldiers dropped off boxes of extra items for the Iraqi patients. At least that was what we could do for the Iraqis besides treating their wounds and injuries. My heart was heavy as I did not yet want to retrograde.

I had to plan to rejoin my unit in Mosul, Iraq as the 86th CSH prepared to retrograde to Fort Campbell through Camp Udairi and then a few other planned places for redeployment. I had mixed emotions about leaving the hospital. I had gained new friends during my attachment to the CSH, both doctors and young soldiers. I met other Filipino non-commissioned officers (NCOs) and young, junior soldiers. While in Camp Udairi, we always joined together for lunch. Mostly, we had white rice, toyo (soy sauce), heated corned beef in the pan, and some days we had Spam or sardines. I remember Butch always invited all the Filipinos into his shop. He worked in the pharmacy. We were in coalition with his boss, and everything was relaxed and comfortable. Ray was the other pharmacist and always offered sodas to drink every time I went to the pharmacy. I praised the Lord for giving me friends such as Butch and Ray. I recollect the days prior to the GAC at Camp Adder. While still in Camp Udairi, Butch, Ray, and I went to the dining facility (DFAC) at 2000 hours, curious as to what the DFAC had to offer. After 2000 hours, the DFAC opened its doors for stop and go, meaning fast-food style dining. One goes inside, gets their food, and eats there or takes it out. It was such a good life. We had hamburgers, beans, green salads, ice cream, soft drinks, fruits, and sometimes ice cream cones or ice cream bars. One night, we saw Chaplain Duke in the DFAC, and he joined the three of us on the west wing of the DFAC. I told him, "Chaplain, I cannot ask any more than this: good food, good company." The chaplain smiled and agreed with me. Chaplain Duke calls us *The Three Catholics.* We were. Chaplain Duke was the hospital chaplain; he is Protestant. One night, I was disappointed as we went to the DFAC. It was such a long line and we decided not to wait and go back to the tent instead. We had only gone there to get some snacks. We found out later that they had served steak and lobster. It was the eve of Easter Sunday, which explained the extra-long waiting line that night. That broke my heart. I happen to like steak and lobster dinners.

I also remember the celebration of the Lenten season in Camp Udairi. It was very solemn as I was living in the middle of the desert. I felt closer to God. The serenity of the night brought a lot of memories to my mind: my broken marriage of 20 years, my son and my family, relatives, and friends back in the Philippines. I thought about the new friends I had and my hope for the future. It was so hard on me, and there came a point when I thought I was done crying. But I cried my heart out until I fell asleep. Crying made me feel better in the morning. As usual, I put a grin on my face but deep inside me was a struggle that I had been carrying. The fact was, I kept postponing my death. That was how I lived all this time: Wait until tomorrow until tomorrow is gone. I realized I need to focus on the mission and move forward. I have to act out my silent freedom into reality and be a hero. It is Easter Sunday, I said to myself, and it will be a new day. It was Easter Sunday, and everything went well. One thing for certain is that I enjoyed working in the hospital because it was safe, and they always had decorations in the halls that made everybody feel at home because those decorations were sent from the United States to Kuwait.

CHALLENGES

One hardship during this deployment was accessing email and Morale, Welfare, and Recreation (MWR) phone calls. There was always a long line to use the computer. In Fort Campbell, I was able to use my computer during my leisure any time I needed and wanted to use it, but this system was outrageous. Sending email had a time limit. I remember when I used the computer to send email at the 86th CSH MWR tent. There was a sign-in log, and then you had to wait for your turn. It worked the same way for making phone calls. I signed in to use the computer and left for the tent. When I came back, to my surprise, my name was crossed out. I was informed that you had to be present when your name was called; therefore, I had to sign in again. Minor things like this were not compatible for my brain, and I worried about it while I tried to fall asleep. To relieve my frustration, I had to get up and walk at least three times around the sleep tents. It was a living hell. I tried my best to deal with it gracefully. The austerity in Iraq should never let me down. Besides, the food in this destitute part of Iraq was not bad at all. We ate tray rations for breakfast and dinner. Meals ready to eat (MRE) were served for lunch. My favorite MREs were the hot dogs and spaghetti. I considered myself lucky if I found an orange- or cherry-flavored beverage mix inside the MRE box because those drink mixes rarely were included. I appreciated the morale barbeque at the CSH. The chicken, ribs, hamburgers, hot dogs, honey beans, and condiments were just fantastic, especially accompanied by watermelon. The food and my silently cherished freedom were the panacea to the MWR frustration the other night.

CSH REDEPLOYMENT

It is time to say goodbye to the CSH staff. I went to a briefing for redeployment in 86th CSH and we were told to take everything slow. We should be patient when driving a car because it has been a while since we were behind the wheel. Relationships developed in theater of war should change. Sex should be taken slowly. Legal matters should be referred to the Judge Advocate for appropriate advice. Intimate relationships with others, with work, and with computers were a part of living an understated life in Iraq—without these things, one falls apart. *Battle buddies* were helpful and helped a soldier from being lonely, away from family and away from home. While waiting for retrograde, I did not know that God sent His angel to pick me up and to finally join my unit in Mosul, Iraq. Bless and praise God! I looked at the angel as Saint Michael, the leader of God's army against the forces of evil. Saint Michael, defender of soldiers in battle, was here to rescue me. God sent him to me on that day camouflaged in an aviator's desert uniform. He asked where he could find a litter patient who was scheduled to be picked up. The other soldier and I looked at each other and were very sure nobody was scheduled to be picked up that day, not even a litter patient. So, I said we didn't have any litter patient for pickup. Yet my silent freedom was screaming like an eagle, "I am willing to be your passenger." He smiled and told me to go ahead, get my things, and meet him at the helipad in 20 minutes. I exclaimed, "Hallelujah! I will be there, sir!" And he left. I told my fellow soldier that I was going back to the tent; she understood and said goodbye. Back in the tent, I shoved nearly everything into my duffle bags and grabbed trash bags to put the rest of my things. I was about to drag my bags with me to the helipad, but once again, merciful as He is, God sent me a hero with the Gator and drove me to the helipad; it was SSG Figueroa, the hero who rescued me from two dozen Iraqi dogs. Tears were welling down from my eyes as we drove toward the helipad. We arrived five minutes early and found no one. SSG Figueroa said that the pilot might be still at lunch, and he was right.

A few minutes later, my Saint Michael showed up and apologized as he was hungry and had to eat. I was so happy to see him, but I kept it to myself and thanked God for the silent freedom. I told him it was all right. After he put gas in the tank, we were ready to go. SSG Figueroa waved goodbye, and I thanked him for all his help, hoping to see him on the other side of the world. We saluted each other and bid our goodbyes. Saint Michael piloted the UH Blackhawk helicopter, and about ten minutes later, we met a dust storm. He told me to hang on as he carefully maneuvered the chopper. With Saint Michael in the cockpit, what did I have to fear? I could have slept, but I was feeling nauseous from the turbulent flight. The journey took a little bit longer, and I felt relieved when he announced our arrival. I told him how relieved I was because I felt sick. He got concerned and jokingly said, "It's a good thing you didn't throw up in my chopper." I smiled and thanked him profusely for rescuing a poor soldier like me. In a saintly manner, he said, "You're welcome. You're one tough soldier, and God will always bless those who believes in Him and will always watch over you." I replied, "Amen." I believed him and I cherished every word he said. With the weapon sling on my chest and with his help, we pulled my duffle bags and put them on the ground in Baghdad Airport. And he left. I stared at the sky until I couldn't see him anymore. It was like seeing a dove flying high in the sky until it disappears from view. It reminded me of the Bible passage in 2 Kings 2, when Elisha was separated from Elijah by a fiery chariot, and Elijah was taken up by a whirlwind into Heaven.

While holding my weapon and carrying my duffle bags on my back and shoulder, I looked for the person manifesting passengers to Mosul, Iraq—my final destination. I almost cried when I saw the first soldier with a Screaming Eagles patch. I recognized him, of course, since he's our IT guy. We exchanged "happy to see you" pleasantries, and he added me to the manifest. It turned out that a Chinook was scheduled to depart with a stopover in Balad to pick up a tank engine. A tank engine? "Hooray!" I replied, as I was very ready to get out of there and fly over Mosul. I took a nap while waiting for the flight and woke up before boarding the Chinook. The Chinook stopped at Balad, and by then it was lunchtime.

We all took lunch inside the Chinook as we waited for the tank engine. I took an MRE out and started cooking. It took about 12 minutes for the MRE to cook. Then, I took an empty plastic bottle out and used my Swiss Army knife to cut off the top and mix my spaghetti. In my canteen, I drank fruit punch. The higher-ranking gentlemen told me I am a survivor. I just smiled, but my silent freedom screamed, "Tell me about it." After waiting for an hour, the pilot told us that the tank engine was not going to show up, and he lifted the Chinook bound to my new home: Mosul, Iraq.

Mosul, Iraq

Nobody was as excited to see me on that day than my commander and the 1SG. I was so excited to see them because at last, I was home with my unit, and it was really good to see the 101st Screaming Eagles again. I suppressed my tears and showed nothing but a brave face. I am an NCO, and I am supposed to be leading troops, so I have to suppress my emotions and keep it within my silent freedom. We have a great commander; he is caring, fair, and brave. His best friend was killed in Kuwait from a grenade assault started by one of the soldiers who had issues about deployment. He threw the grenades over the tents. My soldier and I could have been critically wounded if we slept in the brigade tents after providing personnel records support to the soldiers. But my commander told us to return that night, and so we did. That morning, he grieved so deeply for his friend. He cried, and all I could say was, "Maybe it was his destiny. I will pray for his peace." We both fell silent and prayed for the fallen soldier's soul. During war, anything can happen. I have befriended God, and I gave him my entire mind, body, and soul, and I often prayed for Him to watch over my son.

Aurea on the second level of the Mosul Airport, days after arrival in Mosul, Iraq.

Mosul is 649 miles from Kuwait. It could take about 12 hours by ground assault convoy, on a good day. But it could take days in a hostile environment where the enemy lurks anywhere in Iraq and could deter the convoy. It is equally dangerous when taking a helicopter because we could be hit by an anti-aircraft artillery or surface-to-air missiles. However, it is faster to get to Mosul by air, in spite of the hostile fire. CPT Z said it was a long, dangerous journey from Kuwait to Mosul, Iraq and a miracle to get to Mosul. I responded that it was what actually took place. He said he still cannot believe it because of hostile fires and the fact that no GAC will allow riders outside their units. He was just as glad that I made it safe and sound. He told the others in the division that he was proud of me for being persistent and finally rejoining the group. And then he asked if the dogs chased me again after the frightening experience in Kuwait. I said yes, about a couple more times, but I told them I will shoot if they would not stop chasing me. They all laughed. Then, the ISG showed me my tent and welcomed me in front of the other soldiers. He told me to get some rest and come back to work the next day. After organizing my cot and laying down gear and my rifle next to me, I took a nap and had dinner at the dining facility located next to the D-Rear HQ. How can one rest in a combat zone where no one knows what will happen next? The next day, I was given my mission as the Casualty Liaison Officer serving with the 21st CSH. The 21st CSH made history by being the first Combat Support Hospital to conduct split-based operations during combat with a 164-bed hospital in Balad, Iraq and an 84-bed hospital in Mosul, Iraq, enabling them to provide care for the warriors in an area greater than 250,000 square miles or 160,000,000 acres.

The hospitals provided Level III Combat Health Support, meaning that they are large facilities that take the time to become fully operational but offer much more advanced medical, surgical, and trauma care, similar to a civilian trauma center. The 21st CSH faced the most challenging medical cases and traumas from the two deadliest regions of the conflict in 2003. I felt so proud serving with the CSH and it made me feel that I am really back home with my unit and working with the prestigious 21st CSH. I feel fortunate to serve in two distinguished CSHs that are making history in taking care of our valiant men and women in the battlefield; first, I served in the 86th CSH "Eagle Medics" from 101st Airborne in Kuwait before Saint Michael rescued me. The 86th CSH provided mission command with assigned and attached medical units. It provided health service support capabilities for U.S. and coalition forces and staffed two Troop Medical Clinics, a 25-bed hospital, and damage control surgical teams and other medical capabilities across the Central Command area of operation (AO). I was serving in the 86th CSH when we admitted and treated the first Combat Hero. And then

I served in 21st CSH, 1st Medical Brigade, 13th Sustainment Command (Expeditionary) from Fort Hood, Texas, in the combat zone. This time I told myself, *I've got it made.*

Mosul, northern Iraq, my final destination in OIF 2003.

Bravo Seven

Everything was going fine with our humanitarian efforts until June 12, 2003 when Bravo Seven caught on fire and burned down to only a frame in downtown Mosul. Bravo Seven was our company 1SG's assigned Humvee. The Humvee was utilized for humanitarian missions and on that day, the command sergeant major (CSM) and his driver needed some paints to use for the mission, so they went inside a store to buy some, with the help of an interpreter. Some Iraqis shot RPGs at the city police department. As a result, our CSM and his crew had to clear the area. Bravo Seven failed to start despite trying three times with a jumper cable. Thus, the crew left and decided to come back for it when the area was clear of hostilities. About 45 minutes later, the crew returned and saw what was left of Bravo Seven. The frame stood tall but everything else was completely burnt. Another RPG hit Bravo Seven and ended its utility. It was a sad day for my unit. We were down one vehicle during the GAC for retrograde, but God was on our side on that day as our CSM and his driver came back safely with our rescue team. Several Iraqis were also injured and were transported to the 21st CSH in Mosul.

Special Leave Accrual (SLA)

On July 16, the Chief of Staff General Keane was in Mosul, Iraq to visit the troops and confer with the commanding general. Deputy Secretary of Defense Wolwofitz also came by to witness the life of soldiers in Iraq. Consequently, there was an issue going on. The division CSM was a strong advocate of soldiers and their families. He was fighting about SLA. About 150 soldiers are burdened with more than 90 days of leave in the Army. These soldiers are coming from another hostile field and were consequently assigned to Iraq during OIF. The current policy allows a soldier to only retain 90 days on October 1 (60 regular and 30 SLA). We have 147 soldiers in the division who are well over 90 days—lots of leaders and surprisingly lots of specialists and sergeants, too. Some leaders came to the division this summer with over 90 days because they have had SLA status and have been on back-to-back deployments. When they take leave, it will be all their tax-exempt days. There may be a couple of senior officers and NCOs who were planning on retiring and were affected by this stop-loss, but those numbers are few.

The majority of our soldiers are mid-grade NCOs and officers from 3-101 AVN, 7-101 AVN, 129th CSB, 561st CSB, 2-17 CAV, 626 FSB, 3rd Bde, and 5th SFG. These are the soldiers who have fought in Afghanistan and Iraq, and some were even in Kosovo just prior. I have a Chinook company that returned from Afghanistan and 20 days later was in Kuwait. This caused the soldier to have an excess of 90 days' leave. On September 30, the Army, DOD, and DFAS will take anything over 90 days and reimburse back to 90 with O-5 approval. Soldiers with 112.5 days will gain another 7.5 while deployed, bringing the total up to 120. The soldier will lose 60 days and will be reimbursed 30 days; those lost will be the tax-exempt days.

Here are some potential courses of action:

- Take leave while deployed (but if the soldier couldn't take it in the States, how can he leave his troops during combat?);
- Allow soldiers to voluntarily cash in leave regardless of how many days they have cashed in before;
- Allow soldiers to carry the leave over and give them a final deadline to be at an appropriate amount; or
- Take the leave from the soldiers (bad business).

Again, there are 147 soldiers in the division and Corps Support Group (CSG). I do not have the 5th SFG number at hand, but I'll bet that it isn't pretty. The policy was written before the attack on America's Twin Towers. Everything about America is different now. I have met soldiers with 18 months in the Army who already have two overseas service bars on their sleeve: one bar for every six months in a theater of war.

Big Day

On July 22, Saddam Hussein's two sons were killed in action by the 20th task force in Mosul, Iraq. Uday and Qusay Hussein died inside an old building after a 12-hour gunfight. The soldiers from the 101st Airborne (Air Assault) Division and Special Forces received a tip that the two brothers had been in the area for two months. Soldiers rappelled from a helicopter, blasted the wall, and assaulted the building to catch the two brothers and others. Uday and Qusay were particularly vicious individuals. They were known to be killers of the Iraqi people, using gas chambers or shooting at them with no mercy. This was indeed a big day for the Screaming Eagles of the 101st ABN (AASLT) soldiers. Two days later, it was broadcasted that one of the brothers committed suicide inside the building; but the truth is, they were killed. The Coalition Provisional Authority and Secretary of Defense, Donald H. Rumsfeld, made the decision to release the photos. President Bush was fully informed of the decision, and the Pentagon officials hoped that by releasing the photos, it would reduce the number of attacks against us, the U.S. forces in Iraq.

MOTIVATION

I received my tuition assistance via email from the education center in Fort Campbell, which was expeditious and motivating. I decided to continue taking classes while deployed and as long as there was Internet at the MWR, I could connect with my professor. However, I kept having problems reaching the university. It's a constant battle in Iraq. We were not sure when our email would be read and responded to. Before losing my motivation while waiting for the slow Internet, I decided to go back to the tactical operations center (TOC), a command post. Every day but Sunday, we did battle update briefings (BUB) in the TOC. On some Sundays, soldiers were allowed to relax at the D-Main, the HQ, and take a dip in the swimming pool, but only while practicing safety and ensuring our weapons were accessible in case we got a surprise attack. We had this amenity before the improvised explosive device (IED) became popular in Iraq.

Change of Command

On August 6, 2003, CPT Zasimzcuk had a change of command. For many of us, it was the first time we witnessed a change of command in a theater of war. Anyone who has been in the theater for a while would be happy to retrograde to a place like the United States, where no one is shooting at your rear. CPT Z had mixed feelings, of course; how could he retrograde and leave the troops? But it was time for rotation, so there was nothing else to be done. CPT Z is a great speaker, and all eyes were on him as he delivered his farewell speech. He thanked his lieutenant, LT Cooper, and his 1SG, and also commended the troops on their bravery and loyalty in carrying out their assigned missions. He didn't leave anybody behind, which I know, since I was reunited with the unit. His speech was brief but sincere. He had lost a friend in the theater, but it made him stronger. His loss was our loss. CPT Z was a highly skilled commander, and he was fully respected by all the troops. How many times had he saved my life? My troops and I were saved from the grenade blast because he broke our trip short and told us to get back on the same day. We were supposed to spend the night over the 1st Brigade area and plan to return the next day. We successfully returned to our LSA. The next day, we received a report that SGT Akbar threw a grenade over the tents in 1st Brigade AO. Officers and enlisted personnel were hurt, including the brigade commander. Two officers were killed: CPT Z's friend and another officer who was a major. My thoughts were interrupted by CPT Z's farewell speech as I reflected on the times he spared me from harm.

The new company commander is CPT Arlo Hurst. He came from Divarty in Fort Campbell. He flew all the way up to Iraq to command Bravo Company. The ceremony was a success despite the noisy generator, which had to be fixed in order to hear and understand both commanders' messages.

When looking at the troops in the formation, it was sad to see the number had gone down from 52 to 30 soldiers. They have done so well, but not everybody will be staying in Fort Campbell longer. Some soldiers had to move for another permanent change of station (PCS), others get out of the Army after their expiration of term of service (ETS), some retire if they have met their time and service, and still others receive medical discharge.

Battle Buddy

Romance and temporary relationships are not uncommon during deployment. It was either work or rot. Most people chose work and a little romance or buddy relationship. Everyone needed someone to talk to. I had a battle buddy who was not in the same AO. We communicated by phone and email. The phone calls at nights were relieving and relaxing, a break from the rigors of daily work and worries about families back home. My buddy represented high motivation and courage, a true example of a leader. I looked up to him and so did every young soldier in the division and units that were attached to the division. At least the feeling was mutual. There was confidence and trust. There was a battle of emotions, and there was also a battle of distancing. I played my distance, the right thing to do. It was a doltish situation.

Burn Pits

My work routine helps me focus on my mission. From my LSA, it is about a half-mile walk. I will never forget the burn pits, as I pass them on my way to work. We inhale a massive smoke from these burn pits in our base, and I am told it is hazardous to our health. Yet it is the only route I can take since we did not have much freedom at the time.

I see soldiers add things like feces, trash, batteries, chemicals, heavy metals, and arsenic into the pits. The smell is worse than rotten eggs and stink bombs. But burn pits seemed to be the only remedy to get rid of those things at the time. Later, I found out that burning solid waste may generate many hazardous pollutants and at an even later time, I was told exposure to burn pits becomes a war illness and injury, albeit a hidden one.

Promotion Board Panel

I was a panel member of the promotion board. Taking care of soldiers is a commitment and obligation during times of peace and times of conflict. I had to prepare a minimum of three challenging questions for each assigned topic: communications, map reading, and rules of engagement. A soldier must be tough, able to conquer, and overcome difficulties. I do remember a quote that I read when I was in High School many years ago; Napoleon Bonaparte once said, "Victory belongs to the most persevering." The soldiers appearing for the promotion were prepared, and they all performed well. They were all recommended for promotion and afterward, their fellow soldiers took them to the mess hall to celebrate with near beers and good food. It was the only "drink" we were able to imbibe during deployment. Alcohol was banned because everybody had to be vigilant. After all, we were in a war zone, and we were strictly on a no-alcohol-war-zone policy, under General Order No. 1. General Order No. 1 contains provisions restricting the behavior of troops in the war zone and is intended to show respect to the laws of Saudi Arabia where many troops were deployed. The order also prohibited the possession, manufacture, sale, or consumption of any alcoholic beverage. It also restricted the possession of sexually explicit material. The order influenced those issued in later campaigns, many of which also include bans on alcohol consumption even where U.S. troops are not deployed in Muslim countries.

Shocking News

On August 19, we heard an explosion over the east. We found out shortly that there was an explosion in HQ UN, Baghdad. About 100 individuals were killed. Involved in digging rubbles and finding bodies were 3ID, 4ID, and 101st ABN Div (AASLT). Also, Chemical Ali was arrested and in U.S. custody on August 21, 2003. He was Saddam Hussein's cousin. Ali Hassan Majid is notoriously known by his nickname "Chemical Ali" for using poisonous gas to kill thousands of Kurds in Iraq. He was fifth on the U.S. list of the 55 most-wanted Iraqi figures and was the most powerful member of Saddam Hussein's inner circle still at large during our deployment. His capture was another mission accomplished in this deployment, and the attack was a setback to the UN's humanitarian efforts to help Iraqis who have been through decades of suffering. It was shocking news.

Emergency Leave

On September 8, I went on emergency leave for Marty's funeral, the man who was my step father-in-law. I left Mosul and boarded a C-130 military aircraft. My boss arranged this flight for me so I would be able to leave the country and arrive in time for the funeral. Marty was my step father-in-law with Jane, my real mother-in-law. I arrived in Kuwait with the other soldiers, tired and weary. I left Kuwait the next day and arrived in Amsterdam, then I left Amsterdam bound for Atlanta. The stewardess and the pilot recognized us as soldiers and thanked us for the important job we were doing in Iraq. They even applauded. I felt proud to serve in today's U.S. forces and be part of OEF and OIF, especially in the young troops. What an experience at a young age. I arrived in Atlanta safely and one of the men who worked in Immigration jokingly asked me what I brought for him. I told him all I had was a hot temperature from Iraq and sand from Kuwait. He laughed, though I was being serious in my answer. Then I transferred to a small plane bound for St. Louis, Missouri.

Even after all this efficient travel, I missed Marty's funeral. I came home a day late. Everybody told me that the service was well organized. Afterward, my brother-in-law, Sam, took a group of us to the football game at Busch Stadium. I had to ride the Metro Lane to meet them there. It was one of the best nights in my life, a truly perfect recuperation—seeing a real ball game and having a real beer after seven months of sobriety in Iraq. It was a welcome change to enjoy actual beer after drinking the near beers in Iraq. Though I have to admit that near bears in combat theater are well justified, given the circumstances. Most of all, it was nice to be with people who cared for me. When we arrived home, it was wonderful to see Jane waiting at the door for me. She managed to smile although she had just lost her loving husband Marty. For seven months, I lived and slept in a cot, and it was a positive change to sleep in a real bed with access to a real bathroom in the same building. Life was good. I managed to see my son for five days while in Michigan. He looked well and had developed good manners. I missed him so much, and it meant the world to me when he said he missed me too. My son and I stayed at the new house in Milan where JR, my oldest stepson, lived with his wife Jodie.

While there, I helped JR and Jodie with some yard work. We planted sods, and though it was hard labor, it was also quite relaxing for me. My son took time off to help us. Pat also assisted us by picking up topsoil and sods from the farm. It was such a pleasure laying out the sods around JR's new home. We were doing such a good job that I felt like we could be building homes and doing landscaping as well. We all laughed and enjoyed a few drinks later that day. My son had to go to school the day I left Michigan, so I talked to him on the phone on my way to the airport, and I really enjoyed hearing his voice. I returned to Missouri to pick up my uniform from Jane's house and stayed at Dad Bill and Mom Elva's house in Manchester. They took me sailing on Friday through Saturday evening. They introduced me to their friends, who are all good people. I had a great time laying down on the bow of the boat just reminiscing things. It was good to be away from battle for a while.

I recalled the days in my old home before my husband's betrayal. My son was just nine years old then and did not know I was in town. When his classmates saw me in uniform, I asked them to be quiet because I wanted to surprise CJ. They knew him as Chris, not by his nickname CJ. Only our family circle calls him CJ. The students were able to help with my surprise visit. After the bell rang, everybody went back to their respective classrooms. I had met the teacher, Ms. Chatman, right outside the classroom and requested if I could do something for my son. She agreed to the idea and thanked me for my service. My son was surprised to see me, and he grinned widely. Ms. Chatman introduced me to the class, and I asked if Christopher could come forward to the front. He came up and gave me a big hug. Then I reached for the bronze star from my pocket that was awarded to me. I carefully pinned the medal to my son's shirt and told the class that my son was my inspiration throughout my Army life and my reason for staying alive. I was very proud of him for supporting me wholeheartedly with my job and for being a good student. I saw the pride in my son's eyes, but he didn't allow me to hug or kiss him in front of his classmates.

And I tried not to embarrass him. I was proud to serve with the 101st Airborne Division (AASLT) and to receive the bronze stars. I received two because I deployed twice with the 101st.

I was still reminiscing about the joy in Michigan when I realized I was in Missouri and laying in Dad's yacht. Dad asked how I was doing. I said it was good to be at peace, and he agreed. But though I was physically present in Missouri, my mind at the time was in Iraq. I read in the newspaper about the surrender of an Iraqi general, and I thought it was good news. When it was time to finally return to Iraq, Jane blessed me and so did my other in-laws. Jane noticed that I was limping. She said she had noticed it when I came home. My tight boots caused my limp. I tried to directly exchange them with no success, so I was stuck with those ill-fitting pairs. The pain would not go away many years later. My tight boots deformed my toes, produced blisters between my toes, and aggravated structural problems like a gap in the middle of my foot and bone spurs. It was painful, to say the least. Jane felt sorry for me and told me to stay in touch and take care of myself. I promised her I would.

For me, Iraq was my temporary home. On the return journey, I met two young soldiers in the plane from Amsterdam; one was having a family hardship and could not find a care plan for his four-year-old daughter. His mother could not take care of her because she would be going through a surgery. The other young soldier was fine, and he worked across the airfield from the terminal where I work at. We left Amsterdam and arrived in Kuwait. We reported first to Camp Wolf, then in Tent 14 for manifesting, and finally in the sleep tent. We arrived in Mosul the next day, and I felt at home once we landed in Al Mosul Airport. I found out that our tent now contained a new AC unit, a new shower point, and a new day room with a huge TV. I was surprised to find that they had made so much progress in the two weeks I was gone! On my bunk were two of the four huge boxes I mailed to myself. I went to work and called Kim, my friend from the EO, who was bubbling with excitement and lots of news. I told her I brought her the knickknacks she requested, which made her happy. Also, I gave my boss his welcome back present—tobacco from Cuba. I also left a huge bag of chocolate on the table for my other colleagues. It was good to be back. I sent an email back home that I arrived at my destination safe and sound.

VIP Visit

On September 26, the acting secretary of the Army, the Honorable Brownlee, visited Mosul and briefed the troops. As usual, my secret love was there, leading and guiding the troops. (Yes, I had admired someone secretly, thanks to my silent freedom. I can think what I want to think, and I can admire whom I choose to admire. He was magnificent at his job and he's a great big commander.) After the briefing was over, I had a short talk with one persistent phone battle buddy and softly terminated our relationship. It was for a silly reason. Nobody wants to be rejected; but I did it because I wanted to focus on my mission. An hour or so later, the movie actor Bruce Willis was at the terminal and took pictures with the troops. I got a picture with him, and my fellow soldiers (my boss included) were jealous about it, asking how it was possible for me to get a photo with Bruce Willis in spite of the big crowd. I replied that I just got lucky. Bruce looked tired, skinny, and weary. He put his arm around my waist, and I repeated the gesture so we could pose for our photo op. I didn't feel any fat in his waistline—he was all trim muscle, strong and healthy. He's a classic Hollywood action star. I had the nerve to ask how Demi Moore was doing, and he looked at me with a smile but did not answer.

Later that night, I told my friend Kim about Bruce and my curiosity about Demi. Kim made me feel guilty: "Why did you ask him that question? It might have made Bruce mad and that's why he didn't answer." Bruce and Demi divorced in 2000 after 11 years of marriage. They were one of my favorite couples in Hollywood. After the photo op, Bruce left on a C-130 going back to the America through Kuwait with his band. It was a memorable moment and something I would cherish for a long time.

Freedom Goes to the United States

Freedom takes soldiers on break from Iraq. Some of them say they are going to get married; others are going to a real football game or simply wanting to be reunited with their family. The first 192 troops to return to the United States under the OIF R&R leave program arrived at BWI airport on the morning of September 26, 2003. The group was homogenous in their desert uniform but varied in their personalities and directions for their R&R. For example, one of the soldiers was anticipating his wedding and another was looking forward to his son's birth within the next 48 hours. One was also going home unannounced to surprise his family. A huge welcome celebration came in next to see their soldier for the first time since deployment to Iraq.

Philanthropic Mission

On October 1, we had a convoy through downtown Mosul to visit three primary schools. Annas, our Iraqi interpreter, mentioned that all primary schools in Iraq are public—no private schools. We went through narrow alleys in downtown Mosul, and we were all careful not to run over trenches that could possibly hide IEDs or homemade bombs. There were Iraqi nationals who were happy to see us. We drove in Humvees. The children were very happy to see us. They called me Hijah, a woman who went to Mecca and came back to save them from danger, like a hero. Their smiling faces were so precious. Nevertheless, we were very careful that they didn't get too close to us because they might be carrying a hand grenade or a weapon. As we heard from intelligence sources, the terrorists had been teaching their children and even women on how to handle hand grenades and facts about terrorism. It was a precarious situation. My fellow soldiers told me to get out of the road and pulled my flak vest to protect me as we approached the University of Mosul. Everybody was looking at us, and it was a sticky situation. We didn't know who to trust. The mission for the day was completed, and we returned in the same route to drop off Annas by a bridge. Annas told us that his fellow Iraqis call him traitor because he worked with U.S. troops. I understood his words, and I was concerned about him. I told him to stay safe and not to pay attention to those jealous Iraqis.

Stormy Weather

There was a violent storm on October 2. All the solar panels were down and most soldiers in the tent got wet. Some sleep tents were capsized by the strong wind. It was a big mess. Even one dining facility tent was blown over. The next day, we briefed the brigade commander on mission analysis regarding redeployment. There were three phases: R-30, Transition, and A R + 30. It was a successful briefing and we all learned from it.

Nineveh

On October 8, we went on another humanitarian mission and visited about 15 primary schools in Mosul. It was a long, exhausting day but productive. I saw the great wall in the ancient town of Nineveh, and it was mesmerizing. It's beautiful and detailed with Arabic letters; I wished I understood. Suddenly I felt holy as I realized I was standing on ground where Jonah was buried. I did not take a picture because I was awestruck by it, plus we were moving fast because the area was dangerous—full of IEDs, RPGs, and unexploded ordnances (UXOs). The terrorists don't care where they put these devices, even in a holy place, as long as they incapacitate or kill Americans. Each of us soldiers had a bounty, according to rank: the higher-ranking official they kill, the higher the bounty is. We also passed by the carpet marketplace, which is magnificent. The carpets are beautifully designed and detailed, and the stores carry all sizes of carpets. It would have been nice to stop there too, but again, for safety purposes we had to move quickly to avoid incidents or accidents. The children were so excited to see us, and they all gave us a thumbs up and cheered like we were heroes. They are beautiful children and so innocent. One boy, about 10 years old, was scared to pass us. He demonstrated that we might shoot him. The other soldier and I said no, we're not going to shoot anybody, only the bad guys. We told him he didn't need to worry. With that, he grinned and gave us a thumbs up. We got back late, and I had to go to the LSA to put away all my gear and took a nap. Then it was back to work as usual.

Blinding Dust Storm

On October 15, another storm hit Mosul. I was on my way to the LSA starving and ready to eat when the wind started blowing hard. I had to rush, drop off everything except my weapon (we always had our weapons and ammo) and went back out to return to work. I saw soldiers that turned around because the dust was blinding. It reminded me of Udairi, Kuwait. Once a storm hits, one should just stay in place because it is nearly impossible to see. On this day, I couldn't see anything. It was insane, but I had to continue walking on the gravel until I felt sand. I knew then that I was close to the DFAC. I was starving and I didn't care about the storm. I just wanted to eat. I was fortunate that the DFAC remained open to serve troops. Inside the tent was a second lieutenant who was hit by the center pole that fell down from the strong wind. Her left ear was bleeding, and she was evacuated to the 21st CSH. The guest Airborne girl was in the area at the time and was singing for the troops in the DFAC. However, she stopped once the storm started. The poles were going down one at a time. I had to move to the center because the short poles were intimidating and appeared as if they would hit me at any time. I finished eating my hamburger and hot dogs, then left. About 20 minutes later, the wind stopped and as usual, it looked as if nothing had happened in the area. We were relieved.

One day, I had a mission to go to Q-West with Lieutenant Colonel Whaley, the G-1 at the time. Together with Chief Warrant Officer 3 Negron, we were to fly over in a helicopter. I was in the waiting area for an hour before I was informed that the flight was cancelled due to the strong wind. I was disappointed because it would have been my first time to visit the Q-West, where the Bastogne 1st Brigade was based. This was the same brigade that had the first casualties in Kuwait due to the grenade incident. Anyway, I had to move on and get back to work, update more slides, and participate in the BUB for the day.

RPG Attack

At approximately 2030 on October 16, 2003, we heard four loud booms outside the terminal. We all rushed to don our Kevlar and flak vests. It turned out that 6-7 RPGs were launched at 526th FSB, which is about 10 minutes away from Al Mosul Terminal. Also, there were RPGs launched at us, but they were five to six feet short. This is an unforgettable story to tell the grandchildren, 10 or more years later perhaps. My SPO boss, Major Seelig, told me that this is like FTX, except that this is real, and dangerous. Fun for some people, and fear for others. For me, it was neither one. This is real—no fear, no fun, no hesitation, a rendezvous with destiny. The real report regarding the situation at DREAR came in. The impact areas were 526 FSB and the sugar factory, but so far, no impacts on the airfield.

S- 12 x 107 mm rockets
A- 7x 107MM rockets impacting 526 FSB; 5x 107MM impacted sugar factory
L- fired from LF35621967 (DIVARTY)
U- UNK
T- 162017OCT03
E- 107mm rockets

A Sad Day

On October 17, three were killed in action (KIA) from the incident in Karbala. Lieutenant Colonel Orlando, the Battalion Commander for 716th MP Battalion in Ft. Campbell, KY, and two other soldiers, a staff sergeant and a corporal. They were reported as negotiating with three Iraqi nationals to surrender their weapons. However, the Iraqis started firing at the American soldiers. There were other Iraqis hiding who came out to fire at the three men. It was a sad day. Additionally, seven were wounded in action (WIA). The WIA suffered from lacerations, fragments, and small firearm shots on the face, thighs, back, legs, etc. A few of them were reported serious, but they survived. On October 21, the

Part I: Destination: Mosul

commanding general and his entourage went to Karbala to attend the memorial services for the lieutenant colonel and his two young soldiers. On October 22, a memorial service was offered in D-Rear at DISCOM chapel.

Memories

October 22 is a sad day for me. Amidst war and BUBs, memories came back to me. I married my former husband on this day 20 years ago, a man who would later deceive me. I admit I regret everything. It's sad but true. The only good thing that came out of the marriage is my son. Other than that, my former husband has infuriated me more than ever and I wish I never talked to him since the first day he deceived me. But I trust in the Lord that He may sustain me and give me strength and courage to live life and that I would be able to adjust accordingly. I pray to the Lord every single day that I may meet a man who would care for and love me and those that I care for and love, forever and ever. I pray for a man who shares the same faith and goals that I have; someone I could trust for everything; someone I could trust with my whole life; someone I could trust for a lifetime.

Dahuk, Iraq R&R

Rest and Recuperation (R&R) Day 1. On October 25, together with the other troops scheduled for R&R in Dahuk, we left Mosul by Chinook at 1030 hours and arrived on Dahuk's helicopter pad about 20 minutes later. On that first day, I was glad to meet SPC Nieves from the 21st CSH, along with two guys, Elmquest and Fehlberg. We rode the bus to go downtown. Our guard is from the Kurdish special forces troops and his name is Mike. He translated everything we said. He was also like St. Michael, the Archangel sent from Heaven to protect us from the terrorists looming in the area to kill American soldiers. We combed the downtown area of Mosul and browsed tea sets, scarves, fine china, and carpet. The items were beautiful and relatively inexpensive. I bought some things as Christmas presents for my family.

R&R Day 2. Nieves came by my room early and started chatting with me. She's a live wire, but I'm actually glad she was part of my group. We were a happy group. With the mountains as our backdrop, we had more scenic photo ops, then we went back downtown to Mazi Shopping Center. I found relatively cheap sunglasses, a pair of speakers, and a magic lamp, hoping Genie would come out and give me my three wishes. And then we went to a wonderful place with a huge swimming pool and fountains. The grass and trees were a vibrant green, and a delicious buffet spread was offered all day. The food was just fantastic. We ate vegetables wrapped in a big tortilla, chicken kabobs, and buttered white rice, served with soft drinks, and desserts such as watermelon and honeydew. Hot tea was served last—a special kind with so much sugar that it prevented me from getting sick after overindulging in the local food. It was served in glass shots, and I became fully energized. We started heading back at 1800 hours on the same day and were all alert to stay alive, wary of Iraqi rebels with RPGs. There were several convoy attacks by the enemy recently that cost lives of young, brave soldiers from the 101st and coalition forces in southern Baghdad. We felt fortunate on that day, but would we be fortunate again next time? Only God knows. But the R&R was refreshing and helped recharge our energy for the next day.

R&R Day 3. We drew our weapons at 0900 and turned in all keys by 1200. We boarded three buses and were escorted by the Kurdish special forces as well as American soldiers. It felt good to be doing something for a country. We boarded three Chinooks from Duhok to Mosul. One Chinook had to go back to Duhok to pick up the third batch of soldiers waiting at the helipad. We landed Mosul without accident. However, we were shaken by the sad news that there were about 30 people who were hurt and killed by car bombings in Baghdad.

Ramadan

On October 26, Ramadan will officially begin in Iraq. It is a unique experience to observe this event in a place like Iraq, where Muslims are expected to observe Ramadan. This is the worst period in my deployment in Iraq. It is supposed to be a holy day, but we have been getting more casualties than usual. Each soldier has a bounty. I asked our interpreter how much the bounty was, and he replied $25,000 for enlisted soldiers, and it increases for high-ranking officers. I wish we knew the source of those bounties.

On the second day of Ramadan, the local nationals are showing more violence. This is supposed to be a peaceful holiday. Some people are obstinate and doing something nonsensical—they just want to hurt somebody, period. My silent freedom kept asking, why, why, why? Ramadan is supposed to last for one month. And with the way things have been going since day one, it was very challenging to grasp. On my way to my tent, SPC Bartlett Morris helped me with my bags. He loaded them to his Humvee, and I took the front seat. There was one Iraqi in the back of the truck. I was tired as soon as I reached my tent. What a day. I returned to work at 1530 hours.

Qayyarah West (Q-West)

I went to Qayyarah West, popularly called Q-West by American soldiers, with LTC Whaley, the division G-1. We finally made the trip after it was cancelled numerous times due to bad storms. We picked up the G-1 at D-Main, where the palace is, and it was good to see the palace once again. Our mission was to conduct personnel assistance to our outlying units. There were plenty of personnel actions accomplished in one day. It was three of us: the G-1, CPT Gravitt from G-1, and me. We flew in a UH Blackhawk at 0840 and arrived in Q-West at 0900. We tackled issues with 101st Aviation first and then went to lunch. We went to see the HHC 1st Brigade after lunch. The G-1 had a meeting with the BC. It was the end of the day and time to get back to the camp. It was a cloudy day, but that wasn't a reason to cancel our flight. After all, it was the third time this mission had been scheduled. It was the G-1's turn to drop us off when we returned to Mosul. The chopper touched down at 1700 hours. Another night in paradise. I left my gear as soon as I reached the LSA and had MRE for dinner.

CH-47 Chinook Casualties

Sixteen of 20 U.S. soldiers injured in a deadly helicopter crash west of Baghdad and treated in Germany are listed in stable condition. The CH-47 Chinook helicopter crash killed 16 U.S. soldiers and was the deadliest day for Americans in Iraq since May 1, 2003. The White House said it mourned Sunday's deaths but stated that American resolve was unshakable. Unshakable indeed. We were resolved to successfully accomplish our mission. We were fearless; we were steadfast.

More casualties. November 7, 2003: I'll never forget this day because it's my mother's birthday, and she had passed away when she would have been 93 years old. I reminisced my happy childhood with my parents, four brothers, and three sisters. I am the youngest and spoiled. Oftentimes, my loving mother brought me new clothes and said, "Look what I got for you today, my pretty child." And I ran to give her a big hug. She's the best and I love her so much. I thought, "I will give her everything she wants when I grow up!" My brothers and sisters got jealous, of course, and out of jealousy, I ended up washing the dishes and cleaning the house. But it was all right; I had plenty of new things my mom bought for me. Despite poverty, we were happy. But deep inside me, my silent freedom wanted more—to get out of this rut and buy a lot of things for my mom. I loved my father too, but I loved my mom the most.

In the middle of my meditative remembrance, I was jerked back to the present by the sad news delivered through the Secret Internet Protocol Router Network (SIPERNET) that a Blackhawk from 5/101st AVN was gunned down by an RPG. Four casualties are from the 101st AVN and two others from the other units. We had many casualties two days in a row. These terrorists are demoralizing most of the troops, including me. My silent freedom is screaming, what am I doing here? Maybe waiting for my turn to be the next casualty. But I've been telling everybody that I'll get Saddam Hussein myself. It was a promise, and it would be done on behalf of my brothers and sisters in the coalition forces.

CHALLENGING TIME OFF

I'm resting early tonight. I have a long day ahead of me tomorrow, starting with my school paper, and ending with attendance at evening Mass. That day was also laundry day for me and Kim. But the city power was off, so we went back to our tents. Today wasn't a good day for either of us. We tried to go to KBR and eat lunch, so I took her to my motor pool to dispatch a vehicle. I got the dispatch, but it was locked and somebody else had the key. Instead, we tried to ride the shuttle bus, but it went the wrong way. We got off right at the back of the terminal and walked to the bus parking lot. No bus was there so we had to wait again. About five minutes later, a bus came by—the same bus that we boarded a while ago. Okay, so we ride this bus and then, of all things, the driver decided to take a 10-minute break. Kim and I had had enough, and we got off the bus. Then we saw Major Ferris (the Comptroller in D-Rear). We asked for a vehicle. He gave us his key and we found a truck to drive. Turns out, Kim and I didn't want to mess with a stick shift. We decided, to heck with KBR lunch, and we went to our DFAC near the terminal and had hamburgers instead. Then she talked to another major to fix her keyboard. The major was extremely rude, and she almost cried. I told her to confront the officer and say, "I'm the EOA, and you should know better." We both laughed. Then, to top it all off, we both wanted to wash our clothes, and the power was off. It was not a good day at all. Though I have to admit, I was laughing at myself while writing this story. I thought it was really funny; Kim had just had enough, and we were both exhausted.

Lately, bad things have been happening all around us. There were several casualties from the UH Chinook that crashed in Baghdad. The Chinook had soldiers on board who were returning from their R&R in Qatar, and others were going home. It was a sad day, but the helicopter did not have anyone from the 101st. Last week, a UH-60 Blackhawk was downed by an RPG. It fell somewhere near Tikrit, Saddam's hometown. There were six soldiers who were killed from that incident; four were from 159th AVN, a tenant brigade to 101st Airborne Division. Two were judge advocate general (JAG) personnel, one was a sergeant major, and the other soldier worked for him. Last night, November 15 at 1830, two Blackhawks went down. It was suspected that small arms fires hit one Blackhawk. The Blackhawk's rotor hit the other

Blackhawk, and both fell about 2.2 knots from Mosul camp, by the 3rd Brigade in Rakkasans AO. So far, I received a report with 14 KIA from 101st Aviation Brigade and the Quick Reaction Force (QRF) from 320th Field Artillery. I received another report that three more KIA were from 626th FSB. Four were reported WIA so far. It was a sad day for the 101st. I ate at the KBR with Kim today, and we talked about how sad everything has been in Iraq lately. Even the most motivated soldiers mentioned that they want to go home. It's difficult to motivate anybody at this time, but I did my best and said that everything will be all right. When will these people stop hurting us? There was a report that the oil refinery in Qayrrayah-West was attacked. I worried about my friend in Q-West.

CSM Wilson

On November 23, I was shocked to hear that CSM Wilson from 502nd Infantry Brigade, 101st Airborne Division (AASLT) was killed. Both he and his driver, SPC Ravago, were brutally killed in Mosul by gunshots to the head and a slit in their throats. My silent freedom said these people are savages. It seems like everyone in the Mosul area is turning against us. I was so sad because I had known him and admired him. He was so nice and friendly. Everyone addressed me as master sergeant except him. He just called me by my last name, flatly. He was the only CSM who never wore his badges. I saw several badges on other CSMs' uniforms, so I knew that it was an expected practice. Even though he neglected the badges, I knew he was a well-decorated hero. May God bless his soul and comfort his family and his friends. And may God forgive those local teenagers who brutally killed him, for they know not what they do. We attended his memorial at 502nd and we noticed that it was like a mini battle zone.

It was a dreadful sight. Both streets were blocked by heavily secured Humvees, highly capable soldiers in 101st were ready to defend, tanks were in position, and I can imagine the Apache helicopter pilots must have been on standby in case the enemy made a mistake to attack us at this time. After crossing the street successfully and without incident, we cleared our weapons inside the building by pointing the muzzle down and squeezing the trigger. A bullet was discharged when it was my turn to clear my weapon with the muzzle down. It petrified me. It should not have done that. I could have hurt myself or somebody else. I am glad that our weapons are on the safe mode when not in use. The soldiers helping the small sand box just smiled and told me it was okay. And then we showed honor, respect, and prayers to the Fallen Soldier Battlefield Cross—a symbolic replacement of a cross, which is made up of the soldier's rifle stuck into the floor and the soldier's boots, with their helmet on top. There were two rifles stuck on the ground, one for CSM Wilson and the other for his driver, SPC Ravago. Both soldiers' dog tags were on the rifle and the boots were placed next to each rifle. Then, the Last Roll Call was performed and led by the unit first sergeant. It was a simple yet poignant way for the unit to say a final farewell to CSM Wilson and SPC Ravago. The roll call brought a lot of tears, even to the bravest soul present in that memorial. I brought a lot of tissues to wipe my tears and used all of them during the roll call. After the memorial, we were as cautious in leaving the area as when we arrived, and returned to our LSA, which was just across the road, safe and sound. Every corner was heavily secured, and we were hypervigilant of our vicinity. We were told later on there were actually gunfights in the area and that were suppressed by our brave soldiers. What a disturbing news. The next day, I excused myself from work and talked to one of the lawyers in the Office of the Judge Advocate General.

I explained what happened about the discharged bullet the other night at the memorial. The lawyer said there is nothing to worry about. He said that he was glad to know the system of clearing the weapon works. I slept better that night after hearing the lawyer's confirmation.

The news from CNN on November 23 informed us that three U.S. soldiers were killed Sunday in separate attacks on their military convoys in Iraq. Two soldiers from the 101st Airborne Division were killed Sunday when their convoy came under attack by small arms fire in the northern town of Mosul. After the deaths, witnesses told CNN the soldiers had their throats cut and were stripped of personal effects after they had been shot and wounded while riding in their vehicle. On December 8, 31 soldiers were wounded, including a platoon sergeant and first sergeant. Also, there was a terrorist who blew himself up when approached by two American soldiers. The terrorist pretended to be sick. These situations brought up a G1/S1 conference despite the danger outside the wire. The topic of nearly the entire briefing is redeployment information.

THE CAPTURE OF SADDAM HUSSEIN

On December 14, I was in line waiting for my R&R ride to northern Mosul when we received the news that Saddam Hussein was caught like a rat. Across the Tigris River from his opulent palace, Saddam Hussein emerged from a narrow, dark hole beneath a two-room mud shack on a sheep farm. He was caught in the bottom of a hole with no way to fight back. An Iraqi tipster prompted U.S. soldiers to go to Adwar, about 15 kilometers (9 miles) away. It was early in the summer, and people had to go to the midlevel tipsters and those who had been close to Saddam Hussein. Over the last 10 days or so, the military brought in about 5 to 10 members of these families to help in capturing Saddam Hussein. After they received the actionable intelligence earlier on Saturday, we heard rotors and choppers moving at Mosul Airport, where we were headquartered. The noise told us a raid was about to happen—it became a mission dubbed Operation Red Dawn. Even with reliable information, U.S. forces initially failed to grab Saddam in raids on two different occasions. However, a subsequent cordon and search operation in the same area unearthed the ragged search for Saddam, and we became more determined to get him. Our troops converged on a two-room mud hut squatting between two farmhouses. One room, which appeared to serve as a bedroom, had clothes scattered all over the place. The other room was a crude kitchen. Under the sink, the troops found Mars bars and cans of Spam. On the floor nearby were boxes of rotting oranges. Outside the hovel, the soldiers saw a rug on the ground, pulled it back, and found an inch-thick piece of Styrofoam covering a narrow hole that appeared to be 6 to 8 feet deep. Our soldiers heard noises from below and were about to execute a "clearing procedure" (i.e., firing into the hole or dropping a grenade into it) before the enemy fires back, when someone saw upraised hands belonging to a bearded, bedraggled man. It was Saddam Hussein.

The U.S. forces did not encounter resistance during Red Dawn. Saddam was armed with a pistol but did not struggle during his capture. He seemed to be a man who was tired of running and a man resigned to his fate. When the soldiers assisted the man from the hole, he said in English, "I am Saddam Hussein. I am the president of Iraq. I want to negotiate." The soldiers responded: "President Bush sends his regards." The media coverage was profound and proud of our soldiers' accomplishment. It was indeed a huge mission success!

That was it: Saddam was captured in entirety, and it wasn't the blaze of glory we expected. The troops did not expect to find him where they did. Not only had the troops patrolled past the ramshackle compound before, but they were also taken aback by Saddam's hovel—the messy bedroom and squalid kitchen were a far cry from one of his marble palaces overlooking the Tigris River. What a cover-up. The initial thought from the soldiers was, "Okay, we got his cook." Nevertheless, the troops were glad that their months of tracking Saddam had paid off. During the search, the soldiers also recovered two AK 47 rifles, $750,000 in hundred-dollar bills, and other things. Troops took two other unidentified Iraqis affiliated with Saddam into custody. By 9:15 p.m., Saddam was moved to an undisclosed location by the soldiers and our heroic soldiers continued to guard. I talked with a friend of mine about the consequences after capturing Saddam. Would it stop the war, or is it going to get worse?

CHRISTMAS IN IRAQ

Christmas was celebrated with a live nativity demonstration. I was supposed to be one of the shepherds, but I wasn't able to make it. I was working hard on manifesting soldiers flying out of Military Army AirField and boarding the Strat air. It consisted of soldiers from aviation brigades and some from division support command. After the manifest, I wanted to hear about the midnight Mass. I saw the brigadier general who goes to the chapel every Sunday and asked him about it. He said there was a good turnout for the show at the MWR site and asked if I was part of the live nativity. I replied, "Yes, I was one of the shepherds but was invisible, so you didn't see me." We both laughed. I told him that I was working, and he said he understood. On December 22, there was a Christmas concert. There were five readers, and I was tasked to read the soldier's Christmas. It was a very sad story and I choked up during my performance. But I was told that was how it is meant to be read. I wanted to portray a brave soldier by not crying, but it was very challenging. It was really a sad story. Overall, it was a great midnight Mass, and we sang Christmas carols in between the readings.

On Christmas Day, I had lunch with the soldiers in 311th Military Intelligence inside Kellogg, Brown, & Root, or now the infamous KBR dining facility. We liked it because they cooked good food and served a spread like it's an everyday feast in Iraq. The facility served a delicious spread for the holiday. Turkey, dressing, sweet potatoes, salad, eggnog, near champagne, boiled shrimps, and other tasty treats were served. I had dinner with Kim and MAJ. We watched Farrell Family Reunion in the KBR. It was hilarious. After we finished eating and watching the show, we decided to go back to the LSA. We drove a Humvee, picked up some bottled water, and then they dropped me off at my LSA, my home. It was a nice Christmas away from home. I went to the MWR tent to call my son. There were lots of knickknacks by the door and I took one of the baby wipes and mini bottles of hand sanitizer. Sometimes they have feminine products, and it was truly great to have an outpouring of American supporters who sent us care packages all the time. It alleviated the austerity in our LSA. My thoughts were interrupted when my call went through on the phone. My son's dad answered: "Hello?" I said "Hi" and asked to talk to my son. I didn't want to sound rude to him, but I was still healing from the pain of his betrayal, so I kept the conversation brief and requested to talk to my son. I was so happy when I heard Christopher's voice. He's now 10 years old, and it feels like I missed most of his childhood days. But with the nature of my job, that's how it has to be.

My son did not talk a lot; he was 10 years old, and I missed him so much, of course. To quell my tears and heartache,

I just turned my head to the other direction and kept thinking positive. I told him, "Merry Christmas, my son," and asked how he was doing. I also asked if he received the Christmas packages I sent him. I told him not to worry so much about me and that I will be all right. I told him that God is watching both of us, and God will take me back to him safe and sound. He thanked me for the Christmas gifts and he said he missed me and loved me. That did it all. I love my son so much—he's everything I've got, now that his dad has double-crossed me. His dad had mentioned my son's morning routine after he passed on the phone. Every morning before he left for school, Christopher stood in the front of the TV, hoping to see me in any of the activities that have been going on. He was wishing he would get a glimpse of me at least. It tore my heart. After the phone conversation, I left the MWR tent and went back to the LSA. I was grateful for my silent freedom and wept until I fell asleep. The day after Christmas, all the troops in my unit drove down to the terminal where the Milvans (huge shipping containers) were waiting for customs inspection. It took hours before the inspectors got around to our personal/B bags, looking for war trophies, possible contraband, or things that were prohibited to be shipped.

Convoy to Freedom

It's time for the most anticipated ground assault convoy to freedom. Many of us will return with visible scars of war. But for some of us, our injuries will remain hidden. The roads on the way to freedom are narrow to tread, and I feel like my heart is in my throat as we prepare for ground assault convoy to freedom. I cannot wait to see my son and give him a big hug. As the heavily secured tactical Humvees cautiously positioned themselves at the head of the convoy, we all began to board our assigned vehicle. Both complex training exercises and the real, intensely dangerous world of warfare once again applied in our convoy to freedom. We stopped at Najaf, found bathrooms, and hurried back to the guarded Humvees. Most of us were sweating heavily, but at least it was not a hot blistering day. It was January and about 65°F, which was around the same temperature in December at my home in Honolulu, Hawaii. But I couldn't let my mind wander: There is no time for fantasy. This is real and important, and I had to focus to carefully exit the dangerous, death-defying roads of Iraq together with my fellow soldiers to arrive safely to our destination. The rendezvous with destiny in Iraq was almost over as we drove through the dangerous convoy to freedom. It was finally dark at night and we had to take a short break. A couple of hours to rest our eyes is better than no rest at all. I took advantage of this rest time and tuned out my driver's snore. We slept on our Humvee's hood; others slept inside the van. We're almost there, I said to myself, a few more hours to freedom. I repeated my Holy Rosary and went to sleep.

Then our guard woke us up; it was time to go. We had caffeine and then returned to our Humvees. Bombs might go off, snipers might aim our way, accidents may happen, and close calls are all around us. But we must keep moving, stay hypervigilant for the rest of the convoy, take care of each other, and don't leave anyone behind. I have experienced several close calls in my life during this combat tour. First, I was almost killed by two dozen Iraqi dogs; second, I barely escaped a grenade blast; third, the explosion near our LSA. While in northern Mosul, a fire broke out near our LSA and immediately spread. The wind was blowing away from the LSA at the time; otherwise, our sleep tents would have caught on fire—the same tents that held all our gear. It could have been devastating for all the CSG soldiers. The explosion also shattered the windows of the Al Mosul terminal, letting the air conditioning escape through the holes. The CG instructed us to clear the lobby for possible debris from the explosion. Indeed, there was a piece of metal that landed inside the terminal next to me. Talk about close calls. I waited for a few minutes before touching and investigating the metal, and it almost burnt my hand. It looked like a piece of an M16, black, about six inches long. Fourth, I would have been killed if I did not change my running route during one PT day. I ran every morning in the camp. For safety reasons, we couldn't run in groups even within the perimeter. As I stretched, one of the soldiers said good morning, and I returned the greeting. After stretching, a voice told me to go to the right instead of turning left. I paused, and then I followed the voice. I ran around two miles within the perimeter; it was a good sweat. I went to the LSA and took a good shower, and then walked to the mess hall for breakfast. Another day in paradise, I said to myself. Good morning to you and good morning everybody. Good morning America in the desert. After breakfast, we had the morning brief, and the intel was horrifying. Somebody was run over by a flatbed truck early in the morning, about the same time I did my PT. After they described the exact location of the incident, I almost threw up, and my hands went cold. That was my regular route for my morning PT. It could have been me. My co-workers asked if I was okay, and I said yes, just horrified by the report. The casualty was a female reservist from North Carolina. I bowed and silently prayed for that soldier's soul.

My reminiscence of the past halted as we saw a heavy, expanded mobility tactical convoy with big tanks cautiously heading our way. Who could they be? We all got out and hit the ground, locked and loaded, battle ready. With my silent freedom, I sent the Lord a request: *If this is my time, Lord, I entrust you with my spirit. Please take care of my son.* The heavy front was ready to react. As the other convoy got closer, we were told to stay as we were. It turned out that the oncoming convoy was the Stryker. They had M1A1 Abrams main battle tanks and lots of hardened Humvees. It was great to see fellow brave Americans. It had been a tough year in Iraq. The soldiers waved at us and we waved back, as we all maintained our hypervigilance to our surroundings. We could not afford to be complacent, not when we were in convoy to freedom. Everybody was energetic to return to home sweet home. It was now daytime and thank God for the cool temperature. The good Lord was indeed with us. The first sergeant said we can make a rest stop at the first porta potties we see. About half an hour later, we saw our target, but before everybody ran for it, the first sergeant stopped us and exclaimed, "Ladies and gentlemen, welcome to the free country—welcome to Kuwait!" We all applauded, and everybody screamed like eagles, screeching, "Yes! Yes! Yes! Freedom, freedom finally! Free, free, free! Free at last!" I remembered those words from Dr. Martin Luther King, Jr. "Free at last!" After savoring the air of freedom, we then headed to the bathrooms for relief. It was a long, dangerous, challenging, traumatic, arduous quest to conquer Iraq and to ground assault convoy to freedom. It is hard to believe I had been through a fight, going through a little combat zone to attend a well-guarded memorial in Iraq for our comrades-at-arms who were killed in combat, and rendezvous with destiny. My silent freedom was at the helm, screaming for release, but I had to keep it to myself, whatever it was. I thank my silent freedom, which basically helped me survive through a storm. And at last, I am not home yet, but Kuwait is a stage to freedom. I cannot wait to see my son but was afraid of what I would do with my estranged husband; I cannot live with him. I did not have a home. Again, I bowed my head and prayed silently. God has been with me all this time; He will show me the way.

Conclusion

My story, what I've seen and been part of, is why I would never want to silence freedom. It is vital to my growth and development; it helps me become a better citizen. As I recently reminisced about the first of my four tours in Iraq, I browsed through an article about an American soldier. The American soldier is the face of America; the American soldier is America's might and good will; and the American soldier is the savior of a region unused to democracy. For their uncommon skills and service, for their wits and the choices each one of them has made and those still ahead, for the challenge of defending not only our freedoms but those barely stirring half a world away—these are the reasons why the American soldier was chosen to be TIME's Person of the Year.

Serving in the battleground of Iraq is an honor. We have introduced democracy and it is up to the Iraqi people to safeguard going forward. The American soldier is grateful to those who supported them in times of crisis. The American trust is their strength and stimulus to do heroic deeds without expecting anything in return but a consequence resulting in freedom from Saddam Hussein's tyranny so we can live in a better world.

"Force Package (FP) 1 is complete, 031231. FP2 starts today, 040106 through 040108. Special someone I am silently twitterpated with leaves on Ch 1."

PART II
SECOND DEPLOYMENT.
DESTINATION: TIKRIT, IRAQ

A formation of the 101st Airborne Division, Air Assault (AASLT), Fort Campbell, Kentucky before deployment to Iraq.

OIF-2

I woke up in the morning and found that Contingency Operating Base (COB) Speicher was located on a bed of oil. My fellow soldiers and I were thrilled to see that the news was true: Iraq was rich in oil, and we were sitting right on top of it. After Saudi Arabia, Iraq is the second largest crude oil producer in the Organization of the Petroleum Exporting Countries (OPEC). It holds the world's fifth-largest proven crude oil reserves. Ten percent of Iraqi oil production comes from oil fields in northern Iraq, which are mostly operated by the Kurdistan Regional Government. Ninety percent of it comes from onshore oil fields in the southern part of Iraq, which are under the central government in Baghdad's control. I was so happy to know that COB Speicher was not only a mass of sand in the desert, but also a minefield of oil. To this day, I still reminisce about these happy moments during my second deployment to Iraq.

When we came back from the first deployment in 2004, I remained in the division Personnel Services Battalion (PSB), 101st Airborne Division, Air Assault (AASLT), Fort Campbell,

Kentucky. I found a townhome that was about a 5-minute drive to base. My son stayed with me during this period. I was proud of how he handled my divorce from his dad. I consulted a pastor on base for both of us as part of a well-being check after my divorce. The pastor said that we were both fine and advised further follow-ups until I had fully recovered. Meanwhile, I learned that young children are resilient, as demonstrated by my son's quick adjustment to the situation. I desired a topnotch education for him, and being Catholic, I decided that he would attend a Catholic school. The nearest school was about 20 minutes away from Fort Campbell. That would be quite a stretch, as I had to do PT at 0630 hours and had to be at work at 0900 hours. I asked the base's community service for a referral and received the name of a nice woman who also had children attending the same Catholic school. The Army has every service imaginable and provides a caring community for its soldiers and their families. It is self-sufficient, and all I had to do was ask. If I could stay active duty for life, I would be willing to do so. I am glad that, in 2012, the Army established the Soldier for Life program, the motto of which is "Once a Soldier, Always a Soldier. A Soldier For Life!" The program's objective is to engage and connect the Army, government, and non-governmental organizations to influence the policies, programs, and services that support soldiers, veterans, and their families. The program also aims to build sustainable relationships and results and to reinforce the Soldier for Life mindset throughout a soldier's life.

    I was happy with the new arrangements and explained to my son the logistics of going to his new school, which he accepted wholeheartedly. Being a single parent was not easy. It takes a lot of energy and a leap of faith to simultaneously try to be a perfect mother and a soldier. I was a soldier during the day and a mother at night, cooking

PART II: Second Deployment. Destination: Tikrit, Iraq

dinner, reading books, and assisting with school homework if needed. During a parent–teacher conference at my son's new school, his teacher informed me that he was well-behaved, smart, and doing very well in school. I was proud of my son and his accomplishments at such a young age. Sometimes, I took him to work with me and introduced him to my chain of command and fellow soldiers. Lieutenant Cooper (now a lieutenant colonel in the U.S. Army), who was my boss, said he was very smart and took him to his office to have a man-to-man chat. I really do not know what they talked about, but my son seemed to listen to others better, and he had a lot of friends. I also brought my son to many events, such as the Army Show, where multitalented soldiers demonstrated their abilities with magic, singing, dancing, and comedy skits. I noticed that he liked the show very much. Fort Campbell also has Organization Day, a time for the military families to enjoy amusement rides, games, dunk tanks, a strength-testing high-striker or strongman game, food, and drinks. I was asked to be in a dunk tank but had to say no because of other responsibilities. It would have been fun and a good opportunity to help raise funds for the organization.

Fort Campbell is located along the Kentucky–Tennessee border and is home to the 5th Special Forces Group; the 160th Special Operations Aviation Regiment; and the 101st Airborne Division, the only air assault division of the U.S. Army. The mission on base is to maintain mission-ready forces and to support active and reserve units. The Fort Campbell Parachute Demonstration Team, the Screaming Eagles, contributed to this mission by performing more than 45 parachute demonstrations each year at Fort Campbell and throughout the country to promote the U.S. Army and the 101st Airborne Division. Prior to deployment to OIF-1, I remember seeing them perform one morning as I drove to the base for work. It was exhilarating to watch a rain of brave soldiers jumping from the airplane one by one under blue skies. I had to pinch myself to ensure that it was not my imagination, that I was truly seeing the awesome demonstration of a mass airborne drop right before my eyes. I will never forget the view, and I wish I had had a better camera in my pocket. The demonstration stopped all drivers, who got out of their cars to view the amazing Parachute Demonstration Team at work!

Fort Campbell is in the part of the country known as Upland South and is a beautiful historic area that has so much to offer soldiers, their families and visitors, and those of us lucky enough to call it home. Although the weather might be unpredictable—with snow one day and spring-like conditions the next—one can always count on having something to do while living on or near Fort Campbell, with its parks, rivers, and lakes, as in the picture below; golf courses; a paintball park; shooting ranges; fishing and hunting areas; sports facilities; gyms; bowling alleys; attractions; breweries; and wineries.

Clarksville Marina at Liberty Park.

About 16 miles south of Fort Campbell and the Kentucky border is Clarksville Marina, which offers boat slips, a launch area, and a large parking lot. The amenities include a nearly 2-mile walking trail, a 10-acre fishing pond,

PART II: Second Deployment. Destination: Tikrit, Iraq

pavilions, a playground, a bark park, sports fields, an event center, and many more. In addition to all these activities, my son also learned to play guitar under the tutorship of Butch, a Filipino pharmacist assigned to the 86th CSH, which deployed with the 101st in the previous year. My son did well with his lessons and can now perform on the guitar. My son and I used all the amenities offered on the wonderful Fort Campbell land, including taking swimming classes at the base, playing golf, going boating, having picnics in the park, and many other activities. We also enjoyed going to the Commissary and the PX for food and other things we needed and touring the base.

Eleanor Roosevelt, First Lady and the wife of Franklin D. Roosevelt, 32nd President of the United States, once said, "Life is what you make it. Always has been, always will be." She also said, "The future belongs to those who believe in the beauty of their dreams," and "With the new day comes new strength and new thoughts." I live by these quotations, as they continue to inspire me in rearing my son and preparing him to become successful in life. At the end of the day, I always go back to Eleanor Roosevelt's wisdom and my father's advice, which can be summarized by the well-known maxim: You are what you choose to become.

101ST AIRBORNE DIVISION, AIR ASSAULT

I will forever be proud to be assigned to the 101st and a Soldier for Life. The 101st was the first unit to deploy in support of the U.S. War on Terrorism. Within Afghanistan, the 101st Airborne Division's brigade performed counterinsurgency operations, consisting mostly of raids, ambushes, and patrolling. It also performed combat air assaults throughout the operation. In 2003, Maj. Gen. David H.

Petraeus, or Eagle 6, as the commanding general is known, led the Screaming Eagles to war during the invasion of Iraq, OIF. The division was in V Corps, supporting the 3rd Infantry Division by clearing Iraqi strongpoints that that division had bypassed. Our very own 3rd Battalion, 187th Infantry Brigade, otherwise known as the 3rd Brigade (Rakkasans), was attached to the 3rd Infantry Division and provided the main effort in clearing Saddam International Airport. The division then served as part of the occupation forces of Iraq, with the city of Mosul as their primary base of operations, my first rendezvous with destiny. The 1st and 2nd Battalions, 327th Infantry Regiment, 1st Brigade Bastogne, oversaw the remote Qayyarah Airfield West, 30 miles south of Mosul. The 502nd Infantry Regiment, 2nd Brigade, and 3rd Battalion, 327th Infantry Regiment, were responsible for Mosul itself, which was where I was located on my first combat tour, while the 187th Airborne Infantry Regiment, 3rd Brigade, controlled Tal Afar, just west of Mosul. The 101st also joined in the Battle of Karbala, a city that had been bypassed during the advance on Baghdad, leaving American units to clear the city in two days of street fighting against Iraqi irregular forces. The 101st was supported by the 2nd Battalion, 70th Armor Regiment, along with Charlie Company, 1st Battalion, 41st Infantry Regiment, 1st Armored Division. The 3rd Battalion, 502nd Infantry Regiment, 101st Airborne Division, was awarded a Valorous Unit Award for its combat performance, and after witnessing the soldiers' valor, I was not at all surprised. In fact, I still recall their efforts in going through a landmine. I was at Camp Udairi serving as the casualty liaison officer when reports arrived about a soldier who had stepped on one of the landmines. It was the first such occurrence and a tragic moment for the 502nd and the entire division, as well as for me as the

PART II: Second Deployment. Destination: Tikrit, Iraq

reporter. It is not easy to report casualties, especially if you know the victims. It is traumatic.

On the afternoon of July 22, 2003, the 3rd Battalion, 327th Infantry, 101st Airborne Division, aided by U.S. Special Forces, killed Qusay Hussein and his older brother, Uday, during a raid on a home in the northern Iraqi city of Mosul. The picture below shows soldiers watching as a TOW missile strikes the site of Uday and Qusay Hussein's Mosul hideout during the raid. Soldiers also used Mark 19 automatic grenade launchers, M2 .50 caliber machine guns, and small arms. It was indeed a U.S. tactical victory for the Rakkasans, Special Forces, and the entire 101st Airborne Division.

The Hussein brothers' Mosul hideout on fire during the raid.

101ST RAKKASANS TRANSFERS AUTHORITY TO STRYKERS

In early 2004, the first operational Stryker Brigade replaced the 101st for rest and refit. The existing infantry brigades, artillery brigade, and aviation brigades were reorganized as part of the Army's modular transformation. The Army also activated the 4th Brigade Combat Team, including the 1st and 2nd Battalions, 506th Infantry Regiment, and subordinate units. Both battalions were part of the 101st in Vietnam but saw their colors inactivated during an Army-wide reflagging of combat battalions in the 1980s.

Change of Command

In Mosul, Iraq, our very own combat team, the Bastogne, 1st Brigade, 327th Infantry Regiment, 101st Airborne Division, relinquished control of the Tigris River Valley area of northern Iraq on January 22, 2004, to the 5th Battalion, 20th Infantry Regiment, 3rd Brigade (Stryker), 2nd Infantry Division, in a ceremony at Qayyarah Airfield West. The Bastogne soldiers spent their time trying to improve the quality of life for local Iraqis, capturing people who posed a threat to the lives of both Iraqis and coalition forces. They worked incessantly to find hundreds of weapons and ammunition caches around the area. They built and rebuilt schools, factories, and other parts of the infrastructure. They also helped hold elections in the Tigris River Valley and trained and worked with local soldiers and militia groups. Maj. Gen. Petraeus, Commander of the 101st, stated in his remarks that it was a sign of the great success of the 1st Brigade that they were replaced by a battalion. What the general meant was that the fact that a smaller unit of soldiers was replacing a larger unit signified a job well done. Most crucial problems had already been taken care of and needed to be sustained or improved, as necessary. For comparison, a battalion can have anywhere from 300 to 1,300 personnel. A brigade, on the other hand, consists of between two and five battalions.

PART II: Second Deployment. Destination: Tikrit, Iraq

Colonel Ben Hodges, Commander, Bastogne, 3rd Battalion, 327th Infantry Regiment, and CSM Bart Womack, Brigade CSM, folded the brigade flag to symbolize that they were transferring authority in the Tigris River Valley to the 5th Battalion, 2nd Infantry Regiment, 3rd Brigade (Stryker), 2nd Infantry Division, during a ceremony on January 22, 2004. I remembered these two men from the grenade attack in Kuwait and was glad to know that they had recovered and were back on the battlefield. They are true warriors and leaders who never retreat. Colonel Hodges spoke of the work done jointly by his soldiers and the local government and offered his thoughts on the future. He also shared his belief in the ability of the Stryker soldiers to continue the beneficial work done in the area, to be effective in every facet of operations in Nineveh province where their mission was, and to make history by enabling Iraq to run itself. LTC Karl D. Reed, Task Force 5/20 Commander, expressed his appreciation for the work done by the soldiers of the 101st to make Iraq a better country. He stated that the 101st's accomplishments would forever go down in history as one of the United States' greatest achievements. I thought his last statement was well expressed. I too am proud of the 101st's accomplishments and of being part of and living this history. LTC Reed added that the Stryker soldiers were very excited to accomplish the mission because they were getting a chance to make a difference. Indeed, as I also looked forward to my second tour to OIF, they made a big difference.

Change in Jobs

As I recall, a few months after my first redeployment from OIF with the 101st Division, I requested a reassignment to the 101st Combat Aviation Brigade (CAB), the "Wings of Destiny." I had heard about many successful tactical operations performed by the CAB, and my silent freedom was screaming for my immediate move and looking forward to joining the team. I met with the CAB command sergeant major, my new boss, and my new team, with whom I would be deploying within a few months. At the end of summer 2005, the CAB received its orders to deploy to Iraq. This would be the second deployment for some of us, and we had memorized the drill by heart. My job with the CAB was a little different from the job I performed in the PSB. I would be the non-commissioned officer-in-charge, and with the S-1, we would be leading the CAB's PSB with seven enlisted soldiers, one warrant officer, one first lieutenant, and one major, who was the officer-in-charge, serving more than 3,000 soldiers and aviators in the CAB. The PSB is under the umbrella of the Adjutant General's Corps, which takes care of soldier personnel business, such as finance, personnel service records, correspondence, casualty reporting, evaluation reporting, promotion processes, emergency actions, and numerous other personnel actions. We also initiate paperwork on equal opportunity (EO) actions and serve as a liaison between the brigade and the EO office.

Although my new team's mission included casualty reporting, it was at the brigade level, which is under the division level. The job at the division level was much broader, taking care of the companies, battalions, brigades, and everyone attached to the 101st Division. At the brigade level, my team's focus was on only those soldiers and aviators assigned or attached to the 101st CAB. Thus, I thought it would not be as hectic as my first job at the division level. Hecticness is relative, though, and it truly depends on the situation. If there were a mass-casualty incident, it would obviously strain the health care system and require

PART II: Second Deployment. Destination: Tikrit, Iraq

greater efforts and more emergency units. One can never be complacent on the battlefield.

As we started to prepare for the deployment, I was also busy arranging my family care plan. Despite our divorce, for me, there was no one more qualified and dependable with whom to entrust my son than his own father. He might have made a mistake, but he would take care of our son with all his heart and soul. I felt much closer to my unit, and I decided that the 101st would become my lifelong family, so that it would not be as hard to take after my son left for Michigan. My friends in the 101st were there when I was up, and they were there for me when I was down. At the time, I was deploying to Iraq for the second time with the CAB. It was so exciting to support the valiant pilots and soldiers of the CAB as they resolutely helped move soldiers from one place to another so they could conduct their courageous humanitarian missions.

Wings of Destiny

The Wings of Destiny were my new pride during my second tour to OIF. They were equipped with lethal aircrafts and dauntless pilots. SGT Matson, the public affairs officer (PAO), shared with the S-1 (human resources) team some pictures that revealed his adventures with the 101st CAB, the Wings of Destiny. He showed these pictures, both aerial and ground photos, with discretion, of course. He designed a photo for the newsletter that both he and Sergeant First Class (SFC) Joseph published for the 101st Division. The photos were surrounded by the CAB units' mottos, aircrafts utilized in the battlefield, brave soldiers on humanitarian missions, and on top of it all, a picture of an eagle and the powerful motto "Wings of Destiny." To be a PAO, one must not only be an excellent writer but also have great creativity. I identified the key skills for PAOs: outstanding communication skills, both oral and written; excellent interpersonal skills; exceptional presentation skills; good IT skills; initiative; the ability to prioritize and plan effectively; an awareness of different media agendas; and creativity. Another attribute I wanted to add was courage. PAOs are fearless and willing to cover dangerous raids that could potentially harm them. These descriptions fit SGT Matson perfectly. He was one of a kind. He got along with everybody, and our eyes were always on him as he spoke with excitement, animated by hand gestures. Most of all, we could tell that he loved what he did and was proud of it. When the newsletter came out, we saw SGT Matson's newsletter cover. It was a breathtaking image and perfect for the CAB. I am sure that the brigade commander approved of it; otherwise, it would not have been published. Everyone on my team read the newsletter, and we were awed by the creativity, the pictures, and the write-up. It was exactly what SGT Matson described to us before the publication, and it told a lot about the CAB, its soldiers, and its heroic mission, which cannot be understated. Some photos were deemed confidential, so they were excluded from the newsletter. SGT Matson had been with the Rakkasans on their raids, air assaults, cordons, and searches. The danger he had been through was unspeakable and spoke strongly of his courage, which is why the attributes of a PAO must include courage. SGT Matson decided to leave the CAB after we repatriated to Fort Campbell. I know he liked his job as the 101st CAB PAO, but he had other plans in life. In his newsletter, he had stated that he was leaving after the current deployment to pursue happiness in the land of the free and the home of the brave. Everybody felt sad but happy that he was going for a greener pasture.

PART II: Second Deployment. Destination: Tikrit, Iraq

"Wings of Destiny" Newsletter cover.

The Combat Aviation Brigade, 101st Airborne Division, currently consists of the following units:

- Headquarters and Headquarters Company (HHC), CAB (Hellcats). I was assigned to this unit.
- 1st Battalion, 101st Aviation Regiment (No Mercy), which has six companies: HHC (Avengers), A Company (Spectres), B Company (Bearcats), C Company (Paladins), D Company (Dragonslayers), E Company (Executioners), and B/101 Aviation Regiment (Archangels). They are savers!
- 5th Battalion, 101st Aviation Regiment (Eagle Assault), which also has six companies: HHC (Havoc), A Company (Phoenix), B Company (Lancers), C Company (Phantoms), D Company (Ghostriders), and E Company (Renegades).
- 6th Battalion, 101st Aviation Regiment (Shadow of the Eagle), which has seven companies: HHC (Iron Eagles), A Company (Warlords), B Company (Pachyderms), C Company (Shadow Dustoff), D Company (Witchdoctors), E Company (Trailblazers), and F Company (Sky Masters).
- 2nd Squadron, 17th Cavalry Regiment (Out Front), which also has seven companies: Headquarter and Headquarters Troop (HHT) (Headhunters), A Troop (Annihilators), B Troop (Banshee), C Troop (Condors), D Troop (Dirty Delta), E Troop (Iron Horse), and F Troop (Firehawk).
- 9th Aviation Support Battalion (96th ASB) (Troubleshooters), which has four companies: Headquarters and Support Company (HSC) (Warriors), A Company (Roadrunners), B Company (Big Ugly), and C Company (CIPHER).

The CAB was first organized in July 1968 as an aviation group and is the most decorated aviation unit in the U.S. Army. It was changed to an aviation brigade in 1986. It has served in almost every military operation, such as combat,

PART II: Second Deployment. Destination: Tikrit, Iraq

peacekeeping, and humanitarian aid, since the Vietnam War. In support of the Global War on Terrorism (GWOT), the CAB distinguished itself as the military's premier combat aviation unit during its two deployments to Iraq in 2003 and again in 2005 and during its five deployments to Afghanistan in 2002, 2007, 2010, 2012, and 2015. The brigade flew hundreds of thousands of hours during these combat tours, transporting millions of troops around the battlefield and providing close air support and aerial reconnaissance. The 101st broke its own record for longest air assault in history during the invasion of Iraq in 2003. Previously, the longest air assault was conducted in 1991 during Operation Desert Storm, where my former husband was deployed and fought in the war. He was with the Corps of Engineers. We were both stationed at Fort Benning, Georgia, at the time he deployed in support of Operation Desert Storm. I desperately wanted to serve with him, but the request did not go through, and I stayed behind. I appreciated my silent freedom more and did not say the four-letter swear word. When he came back, it took weeks for him to settle down and to forget about jumping into foxholes and facing vicious RPGs. The aftermath of war is brutal. There is no assurance that a solider will heal from the trauma of war. My former husband eventually realized that he was back in the United States and that the war was over. I finally saw where he had been in the gulf when I deployed to OIF and saw ruins caused by relentless bombings in Iraq. I prayed for the souls of the valiant soldiers who did not return home and those wounded during the war.

As my mind went back to my own deployment, I appreciated my assignment with the 101st CAB and admired and trusted its valiant men and women aviators and soldiers in the face of danger. The CAB departed Fort Campbell and went straight to Kuwait. During the three- to four-week wait for the flight to Iraq, soldiers occupied large open-bay tents and had to acclimatize and adapt to the heat in Kuwait. During a previous tour, I had become familiar with the vicinity: the muggy weather, with high temperatures of 150°F; the sand; the little bazaars; the hadji shops; the training; and the support. I also thanked God for the chapel where we could pray for God's blessings. I noticed the improvements in the area; for example, each tent had an air conditioner and was bigger, so we had more space for additional cots. However, we still had to walk about a quarter of a mile from the living area to the dining facility (DFAC). It felt like it took forever to get to the dining facility, probably because of the humidity. Soldiers used soft caps and always had their sleeves down to prevent sunburn. But the long walk paid off upon reaching the DFAC and the smell of the food. At times, the DFAC served steak, meatloaf, fried chicken, vegetables, ice cream, a good selection of pies, and much more. I was cognizant of how expensive it was to take care of deployed men and women at war. In return, as soldiers, we put forth our best effort for mission success. A month later, the advance party departed for Iraq. Destination: COB Speicher. Then followed the remaining party, including me. The advance party welcomed us in a dark and quiet place; after all, we were at war, and we did not want to wake up the enemy lurking to kill American soldiers. According to a Congressional Research Service report, there were 143,800 boots on the ground in Iraq in FY2005.

In August 2005, there was a ribbon-cutting ceremony at Forward Operating Base (FOB) Speicher, which was renamed as a contingency operating base to welcome the new coalition headquarters at the site. This was the largest structure built to date for coalition forces. In November 2005, the Division Headquarters, 101st CAB, and 1st and 3rd Brigade Combat Teams (BCTs) deployed to Iraq for a second time. I deployed with the 101st Division the first time and with the CAB the second time. As Task Force Band of Brothers, the division assumed responsibility for the northern half of Iraq, which was the largest area of operations. Under the new modular structure,

PART II: Second Deployment. Destination: Tikrit, Iraq

the 2nd and 4th BCTs and the 159th Combat Aviation Brigade were attached to other multinational division or multinational force commands in other areas of Iraq. The majority of the division was to be deployed in late 2007, two years after my deployment to COB Speicher, Tikrit, Iraq. The division's 1st, 2nd, and 3rd BCTs and elements of the sustainment brigade deployed independently to Iraq, where each served under the command of different multinational divisions then conducting combat operations throughout Iraq.

The 101st CAB occupied Al Sahra Airfield, an air installation near Tikrit in northern Iraq. The installation is approximately 106 miles north of Baghdad and 6.8 miles west of the Tigris River. Based on the description given in Genesis 2:8–14, many Bible analyses conclude that the site of the Garden of Eden, in the Middle East, was somewhere near where the Tigris and Euphrates Rivers are today. Together, these two great rivers define the ancient region of Mesopotamia, which comprises present-day Iraq, as well as parts of Iran, Syria, and Turkey. According to Genesis 2:10–14, four rivers were associated with the Garden of Eden, namely, Pishon, Gihon, Hiddekel (now Tigris), and Phirat (the Euphrates). I was amazed that we were seriously engaged in war in a place where civilization began. There should be peace, but instead, we were fighting a war in the place where Adam and Eve were created. Time had changed, but not the memories of 9/11, which resulted in multiple OIF deployments for justice. Soldiers were making a huge difference in support of OIF, sacrificing their lives so that people could be free. Now that Saddam Hussein was gone, we were hoping that there would be peace in Iraq. However, an insurgency emerged to oppose us and to try to suppress the Iraqi forces we were training to sustain the defense of the Iraqi people.

The Al Sahra Airfield is served by two main runways measuring 9,600 feet, with a shorter runway measuring 7,200 feet. The Americans originally christened the installation as

Forward Logistics Base (FLB) Sycamore, but the name was later changed to Forward Operating Base Speicher and then changed again to Contingency Operating Base Speicher. COB Speicher was named after Captain (CPT) Michael Scott Speicher, a U.S. Navy pilot who was killed in action in Iraq during the 1991 Gulf War, in which my former husband also served as a warrant officer. CPT Speicher was born in Kansas City, Missouri, and then his family moved to Jacksonville, Florida, when he was 15 years old. He became the first casualty of the Persian Gulf War on January 17, 1991, when his plane was shot down over Iraq. However, based on findings near the wreckage of CPT Speicher's F/A-18 Hornet fighter, the U.S. intelligence agencies obtained information indicating that CPT Speicher might have survived the crash and been held in captivity in Iraq. At that time, the U.S. Navy changed CPT Speicher's status from "killed in action" to "missing in action." After years of investigations into CPT Speicher's fate, the Navy reported that his remains had been found in Iraq and later returned to his hometown of Jacksonville.

CONTAINERIZED HOUSING UNIT (CHU)

Our living quarters, known as the chu, COB Speicher, Iraq, OIF 2005–2006.

PART II: Second Deployment. Destination: Tikrit, Iraq

Our logistics support area (LSA) was surrounded by a T-wall, a 12-foot-tall steel-reinforced blast wall. Before the deployment, the barracks once housed the students of the Iraqi Air Force Academy. Some of the quarters were small and looked like one-level modern houses in a row of about 10, similar to condos. They were renovated before we arrived and supplemented with specially outfitted shipping containers. We called them chus, for containerized housing units. The priority for the U.S. ordnance team was to clear out the tons of unexploded ordnance in the area. We occupied the chus when it was safe, and we were assigned our roommates, who were from the same unit and of the same gender. I lived in the first chu facing north, with a female soldier from the EO office as my roommate. The chus were all wired for electricity and air conditioning for the hot summer, and many soldiers had their own satellite TVs. I did not get one because I did not need one. I focused my attention on finishing my studies online and watched TV at the Morale, Welfare, and Recreation (MWR) tent. I like to have other people around that I can laugh with or cry with, rather than watching a movie by myself. It is too dangerous to feel alone in the battlefield. My roommate and I had different work schedules, so most of the time, I was alone, as was she. We got along well and organized our beds, put a curtain between us for privacy, always cleaned our small space, and made our quarters look more like a home. Our mail arrived daily, as compared to only two or three times a week at outlying base camps. For me, it was living freedom compared with my first deployment, where soldiers lived in tents in bay areas with no running water. At COB Speicher, the bathroom was a separate building located in front of the chus. With running water, everything felt like a luxury. We had both hot and cold water, so the soldiers were very excited. However, Commander Gina noticed the soldiers' wasteful habits, so she instructed the first sergeant (1SG) to implement basic energy conservation rules. We were told to turn off the lights if not needed. We were also instructed not to waste water: When brushing teeth, turn off the water, and then restart it when rinsing. The same rule applied when taking a shower: Turn off the water when not needed, and turn it back on when rinsing. The energy conservation discipline from the Tikrit deployment is very much alive in my home to this day, as I practice it every single day and try to pass it down to my kids. It saves a few dollars while also conserving energy.

Our worksite was about 50 meters from the chus. It was also surrounded by T-walls to protect us from enemy attacks and the persistent dust storms. The S-1 team was co-located with the Logistics Office, the Public Affairs Office, and the EO Office where my roommate worked. SGT Matson, the PAO, worked late nights to finish his news reports, as did SFC Joseph, his supervisor. Being located in the same building had its advantages. With these strong offices by our side, we had more leverage than other soldiers, for example, in not running out of office supplies and in getting firsthand information about the news before everyone else did. In addition, the IT office was located right next door, so we could just walk into their office to request technical assistance and obtain a quick resolution. There was no need to drive.

A small PX was established nearby, about 50 meters from the chus, and it was convenient to shop for Slim Jim beef jerky, popcorn, and other basic needs after work. The fairly large main PX was located in the opposite direction from the CAB, at a distance of about two to three miles. That would mean a maximum of six miles just to go to the PX and back. The store served about 2,000 soldiers a day. Our legs were our main vehicle during the deployment. The constant walking helped balance the calories we took in from the DFAC and made us stronger. Later on, I suffered from a bad back, bad knees, and deformed feet. My foot problem was aggravated by my multiple deployments and with my silent freedom, but as the saying goes, I just sucked it up.

In the main PX, I noticed a freezer section where frozen meats shipped directly from Germany were sold. I might as well have worked in the Contracting Office, as I became curious about where everything was originating from, including the meats. The U.S. soldiers, Department of Defense (DOD) civilians, and contractors had access to various amenities during the deployment, including fast-food restaurants such as Subway, Burger King, and Pizza Hut. They also frequented the coffee shop, which was similar to the popular Starbucks in the United States. Some of the soldiers expressed a desire to get into owning their own coffee shops after we repatriated to the United States. Also next to the PX were a beauty salon and spa, a barber shop, a dry cleaner that also provided alterations, a photo lab, and a gift shop. The cinnamon bun shop was exceptional. We were not sure how the owner did it, but soldiers flocked to the shop. One Saturday, soldiers were brokenhearted when the shop was closed as a result of a power outage. Soldiers patiently waited outside just to have a cinnamon bun. A contractor paid from his own pocket for all of the services that the beauty salon provided for the soldiers that same day. It was amazing how people could be very generous in the war zone.

I was grateful that the main PX was not very close, so I was not tempted to shop every day. However, I did buy a couple of golden cartouches with my name on one side and the word "Faith" on the other. I felt different when I tried on my new cartouche, as if I had worn one before, but a more elaborate one. Again, my karma was kicking in. A cartouche is oval in shape with a line at one end at right angles to the oval, indicating that the text enclosed is a royal name. I wish I knew my name from my other life. I also bought a 24-karat gold bracelet that I have worn only once and is collecting dust in my drawer.

Other Facilities

COB Speicher had a 32-bed hospital that was supported by the 28th CSH from Fort Bragg, North Carolina. The chapel was also nearby, and Major Kopek, the priest, would become my neighbor a year later after redeployment to Fort Campbell, Kentucky. I felt blessed and secure having a chapel within the brigade. My worries and anxiety disappeared whenever I went to the chapel and attended Bible study. I became a lector coordinator in the Catholic services and joined the choir. It was okay for me to join the choir as long as they did not ask me to sing solo. I thought that it might drive people away if I sang solo and the priest would blame me for it. On a serious note, it was overwhelming to discover many vocal talents in the 101st CAB. There was no problem finding choir members, and I felt that the Holy Spirit was always on our side because of these extraordinary voices.

MWR Facilities

Kellogg Brown & Root (KBR) was a contractor who ran two MWR facilities at COB Speicher. They installed Internet terminals, phones, big-screen TVs, PlayStation 2s, pool tables, table tennis tables, and libraries and built a lounge area. I heard about a few movie theaters scattered around COB Speicher. They were giant TVs where soldiers played movies. It was good entertainment, if there was nothing else to do. One of the MWR tents was within walking distance from my LSA. This was exciting to know because I needed to use the computer to finish classes online. The connectivity was much better at the MWR tents or sometimes at the office after work. I never stopped taking classes, even during deployment. The classes kept me engaged and helped me maintain my sanity.

PART II: Second Deployment. Destination: Tikrit, Iraq

Everything was within walking distance from the COB, including the DFAC, which was about 700 meters from the chus. I was told that, before our deployment to COB Speicher, the 1st Infantry Division Support Command (DISCOM) celebrated the grand opening of the Victory Inn at FOB Speicher (now called COB Speicher). The food services team made their dining facility a world-class operation, serving the best meals in the country for soldiers and deployed civilians. They served delicious meals with prime rib, crab legs, and fried rice. For dessert, the soldiers and partnered contractor, KBR, served two beautifully decorated cakes complete with the 1st Infantry Division logo. The chefs also decorated the dining facility with edible decorations. It was a considerable effort and a huge success. We were grateful as we heard the story, because they made the facility suitable for our deployment. The dessert bar was sinful, as they offered four different types of Baskin-Robbins ice cream. They also made milkshakes and had several fruits and mixes to choose from. The chefs showed their creativity by having a different theme each week. One week, they prepared Mexican foods; the following weeks were Italian, Asian, American, and so on. Pizza night became my favorite, after all the healthy foods served during the week. It was easy and more convenient to take the pizza out for dinner later. Snacks such as cookies, pies, cakes, and many others were also readily available. Oftentimes, our team members took turns bringing desserts to the office to snack on. We were like a big family away from home. The DFAC remained open for midnight chow, and sometimes, it was challenging to resist. We gained a little weight during this deployment, and we all blamed it on the chefs and the delicious food they served. The soldiers at COB Speicher had it made compared to the soldiers who lived "outside the wire," or beyond the base security perimeter. We were allowed to wear soft caps on base, although Kevlar and full "battle rattle" uniforms had to be worn outside the wire.

Dust Storm

Tikrit is in the middle of a desert. There are no trees, no plants, and no greens. The only living creatures I saw were us, the human beings; the scorpions that inhabited the moving sand—mysterious desert insects that resemble the back end of a camel spider but also have pincers in front and a long tail that curves over their backs; and other crawling, creepy bugs. Some soldiers joked that it was no wonder Saddam Hussein did not want to live in Tikrit, as it is as miserable as Hell: located in the middle of the desert and hot and muggy, with temperatures reaching a soaring 130°F in summer. (During the winter, it does get cooler, reaching a surprising 65°F in December.) We all laughed at the joke, although it had some truth to it. Dust storms are also common in Tikrit, and we saw 1,000-foot-high walls of dust. Nobody in their right minds would want to fly in during a dust storm, not even my Saint Michael, who piloted a Black Hawk during my first deployment from Camp Udairi, Iraq, to the Green Zone in Baghdad. High winds last for hours or even days, which pretty much shut down our operations. The winds are so powerful that they have the strength to pick up anything, including helicopters, and move them a few feet, causing damage.

I asked one of the pilots if they fly in hazy weather. The pilot told me that a bad weather flight is certainly possible when there is no lightning. Pilots delay a flight if they know that lightning is a possibility. In many cases, a pilot can land the chopper safely even if it is hit by lightning, but the safest way is to delay the flight, especially in the battlefield. There is no secure land in Iraq except in the airfields occupied by

American troops. Also, just because a helicopter lands and the passengers are safe does not necessarily mean the helicopter will be damage-free or safe. Lightning and bad weather can cause significant damage to the helicopter, which can be costly to repair. He added that it definitely is not wise to fly when a dust storm is anticipated. It is suicidal.

An Apache helicopter in hazy weather with no anticipated lightning.

FLOOD IN TIKRIT

On February 5, 2006, continuous rainfall flooded our office in Tikrit. Our brigade S-1 officer, Maj. Hargrow, and Chief Smock led us to the office to save our computers. Our building had six steps and was on higher ground. It was dark and dangerous, so we all used flashlights and avoided any floating objects. After securing our laptops, SFC Schroeder and Staff Sergeant (SSG) Warren rescued us and took us to a dry place. SFC Schroeder scooped up Maj.

Hargrow first and then came back to transport me to the dry area. SSG Warren scooped up the lieutenant, who also joined us in the dry area. The Chief and SGT Broadnak rescued the rest of the S-1 crew until everyone was safe. These men were our strongmen who saved us one by one. The water came up to about the waistline of SFC Schroeder, who is about 6 feet tall. I am 5'3", and I assumed that the water would be at my neckline. I can imagine that I would have been swimming in the water if not for the timely rescue. These men were my heroes in Tikrit. Later on, 1SG Kolb ensured that we were all safe and that all of the rescued equipment was dry and functional. We helped reorganize the supply room as soon as it had dried out. This situation led me to remember the floods in the Philippines. The climate in the Philippines is divided into two seasons: rainy or monsoon season, from June to the early part of October (sometimes even through November), and dry season, from the later part of October or sometimes December to May. We lived through floods every monsoon season, often evacuating to higher ground or a safer place until the ground dried out. I remember staying safe on the second level of our house until the water subsided, and my parents had to rebuild our first floor. This became a routine for my family, since it flooded every year. When Specialist (SPC) Parker called my name, it snapped me out of my memories about other floods I had experienced.

BED OF OIL

The February 2006 flood in Iraq also led to the displacement of about 400 people in the suburbs of Baghdad. In Mosul, U.S. forces helped move people to safer locations. In Tikrit, the Iraqi Army improvised a camp to house flood victims. The rain finally stopped after three days, and everything began to dry out, so the S-1 officer led us to the office and

PART II: Second Deployment. Destination: Tikrit, Iraq

investigated which equipment needed to be repaired. We were amazed at what we were walking on: a bed of crude oil! It was exhilarating to be standing on a bed of oil, and we were all laughing, happy about our newfound discovery and celebrating the joyful moment. We were flooded, so the crude oil was all over the place, including the outskirts of the dining facility. Crude oil is unrefined petroleum composed of hydrocarbon deposits and other organic material. It is a slippery, viscous liquid that is dark in color and not miscible with water. It is a natural resource that is extracted from the earth and can be refined into products such as gasoline, jet fuel, and other petroleum products.

Oil looks the same, of course, no matter where it is found, but the fact that we were standing on a place with abundant oil was just unfathomable. This story that I had been hearing since OEF/OIF had started was not a hoax after all. I thought it was insane for troops to be deployed and to sacrifice their lives largely because of oil. Nevertheless, the news reported that General John Abizaid, former head of U.S. Central Command and Military Operations in Iraq in 2007, later stated that, "Of course it's about oil; we can't really deny that."[1] Former Federal Reserve Chairman Alan Greenspan agreed, writing in his memoir, "I am saddened that it is politically inconvenient to acknowledge what everyone knows: the Iraq war is largely about oil." My silent freedom was shackled; I was shocked.

THE REASON WE ARE AT WAR

On the morning of Tuesday, September 11, 2001, the United States was struck by four coordinated attacks

---

[1] Antonia Juhasz, "Why the war in Iraq was fought for Big Oil," *CNN*, April 15, 2013, https://www.cnn.com/2013/03/19/opinion/iraq-war-oil-juhasz/index.html.

by the Islamic extremist group Al-Qaeda. Nineteen militants associated with this terrorist organization hijacked four airplanes and carried out suicide attacks against targets in the United States. Two of the planes were flown into the twin towers of the World Trade Center in New York City. A third plane, American Airlines Flight 77, hit the first floor of the west wing of the Pentagon. Upon Flight 77's impact, all 58 passengers, crew members, and hijackers died instantly. Inside the building, the dead numbered 125, including a friend of mine who was stationed at the Pentagon at that time. Their bodies were burned, and only their skeletons remained, in seated positions like they were having a meeting. The fourth plane crashed in a field in Shanksville, Pennsylvania. The attacks killed almost 3,000 people and left more than 25,000 injured, with many facing substantial long-term health consequences, in addition to causing at least $10 billion in infrastructure and property damage. It was the single deadliest terrorist attack in human history and the single deadliest incident for firefighters and law enforcement officers in the history of the United States. It was also the reason for major U.S. initiatives to combat terrorism and a series of U.S. deployments to Iraq. How can we forget?

The reason we went to war was to find and stop the alleged development of WMD and to end, once and for all, the purported link between Saddam Hussein's government and terrorist organizations, particularly Al-Qaeda. In other words, OIF was part of the broader GWOT; it was not about oil. As far as I remember, in his speech on March 19, 2003, President Bush addressed the nation and outlined the purpose of invading Iraq: "to disarm Iraq, to free its people, and to defend the world from grave danger."

I also recall President Bush's speech when he met with troops in Baghdad, Iraq, in 2003. He said, "I bring a message on behalf of America: We thank you for your service, we're proud of you, and America stands solidly behind you. Together, you and I have taken an oath to defend our country. You're honoring that oath. ... You are defeating the terrorists here in Iraq, so that we don't have to face them in our own country. You're defeating Saddam's henchmen, so that the people of Iraq can live in peace and freedom. ... By helping to build a peaceful and democratic country in the heart of the Middle East, you are defending the American people from danger, and we are grateful. You're engaged in a difficult mission. ... We did not charge hundreds of miles into the heart of Iraq, pay a bitter cost in casualties, defeat a brutal dictator, and liberate 25 million people only to retreat before a band of thugs and assassins."[2]

It was a very inspiring speech, and it should be remembered that way. The thought that we were fighting for oil was difficult to comprehend, and I still wanted to believe that we were fighting for a better cause, like making a big difference in freeing the Iraqis from tyranny, finding the missing WMD, and helping successfully end the GWOT, and not fighting for oil. It must have been an expensive idea. I was sad the next few days about my discovery. I kept thinking about it before I went to bed and when I woke up. Even when going to the church, my mind was somewhere else. On my way to the dining facility, I still thought about it. It even slowed me down at work. Was it all worth it? The oil of Iraq cannot be paid for by a 27-mile road march nor by a 9-mile run. We lost thousands of soldiers—KIA, WIA, and non-combatant deaths—in this war. All these efforts were in support of OIF and not because of oil. Nevertheless, I had to learn to let it go. I have learned to think of a more positive justification for our presence in Iraq.

---

[2] Office of the Press Secretary, November 27, 2003.

An armed Observation Helicopter (OH)-58D Kiowa Warrior on reconnaissance and security patrol flying over the city of Mosul.

Live-Fire Training

I could not have been prouder than I was being with the 101st CAB in the battlefield. They are authentic and generous warriors. They are brave and always prepared to face the risk of battle. They are the Wings of Destiny, flying our heroes to the world of terrorism and helping people find peace and freedom. My team flew to the towns of famous Taji, Al Asad, Balad, and other neighboring towns in Iraq and provided personnel support to the soldiers. I had to suppress my excitement, as I had been wanting to see Taji and our soldiers assigned there. The food in Taji was good, although COB Speicher's was better. On our way back to COB Speicher, we got a great surprise from the aviators. It was an exciting trip flying in a Black Hawk over the now-populated Babylon. I could not hear what the pilots were saying, so I had to wear a headset. They were talking about firing our M4s as part of training, and the co-pilot said now was the time. The live-fire training was unforgettable for me; we were firing live ammunition as the helicopter tipped its nose downward. We

PART II: Second Deployment. Destination: Tikrit, Iraq

were at high altitude, of course, and SGT Broadnak and I were excited and relayed the story over and over to our S-1 officer and the team after we landed at COB Speicher. It was awesome. I slept well that night, both because of the trip and from the grand live-fire training. It was a journey that I will remember for a long, long time.

Marksmanship

Another training exercise in Iraq was weapons qualification, or marksmanship. Qualifying in marksmanship in the desert was not as cool as qualifying at Fort Campbell, mainly because it was hot and a different environment. By regulation, we are supposed to conduct marksmanship training biennially, and this was not an exception. We were driven in Humvees about 30 minutes away from the LSA at COB Speicher and fired our weapons there. Unlike during the first deployment to OIF, we had the time to train. I had missed the weapons marksmanship back then, so I was glad we did it. This was one of the moments when I appreciated my assignment with the 101st CAB. Things made sense, and everyone was pleasant. I was assigned to the HHC, which had about 100 soldiers. Our 1SG Kolb was a gentleman; he was seasoned and a good NCO. He was responsible with his decisions and respected each soldier, treating them with dignity yet with fairness. The soldiers liked him because of those attributes. As first sergeant, he had to be tough, and he showed that attribute too, if a soldier challenged his orders. After the marksmanship training, we headed back to the chus and cleaned our weapons. We were all issued M4/M4A1 5.56-mm carbines, which most soldiers really like and prefer over the old M16A2 rifle. The M4A1 is a lightweight, gas-operated, air-cooled, magazine-fed, selective-rate, shoulder-fired weapon with a collapsible stock. It has replaced the

M16A2 rifle and is now the standard-issue firearm for most units in the U.S. military. The short barrel, collapsible stock, and detachable carrying handle with a built-in accessory rail provide soldiers operating in close quarters with improved handling and the capability to rapidly and accurately engage targets at extended range, both during the day and at night. The M4 carbine shares more than 80% of its design with the M16A2 rifle and replaces all M3 .45 caliber submachine guns, as well as selected M9 pistols and M16 rifles. The M4 series is a shortened variant of the M16A2 rifle.

Increasing Insurgency

Before the 101st CAB took over, the 1st Battalion, 10th Aviation Regiment Dragons had logged over 17,000 supporting missions overseeing several areas in Iraq, including Mosul, a city I knew from my first deployment to Iraq in 2003. The city of Mosul was a hotbed of insurgency in 2003 that took the lives of most of our soldiers, including the fierce CSM Wilson and his driver. I will never forget when our lieutenant broke the sad news early in the morning on that day. I was just waking up when she said that CSM Wilson had been killed. I gasped in disbelief and burst into tears. I had just spoken with him the day before and still remembered him saying that nobody wanted to mess with him because he was brave. I believed him because he was truly courageous. During the briefing, our intelligence officer stated that a group of Iraqi teenagers threw a rock onto the windshield of the vehicle the CSM and his driver were in and slit their throats before they could draw their weapons. It was indeed very depressing news, as we felt that we lost our right arm in the battlefield. A few months before we departed Mosul, the enemy created the magnetized IED, also known as a sticky bomb, a simply constructed yet very lethal device. Magnetized IEDs caused more casualties, but our intelligence

PART II: Second Deployment. Destination: Tikrit, Iraq

officer reported a countermeasure to detonate those sticky bombs before they exploded. The insurgency was increasing as our redeployment came closer, and we were briefed that the Iraqis did not want us to leave. This was understandable because they would be losing a major source of their economic activity. War is not a cheap undertaking; the costs of planning and execution are greater than we could imagine and had helped the economy of the Iraqis. However, as soon as the coalition forces departed, the radical leftist Iraqis tore down the infrastructure that we had helped build and bombed their neighborhoods, dragging the country back to the conditions in which we had found it. It was sad, but we could only do so much, and we had to start building again, starting in our very own backyard, our home, the United States.

LETHAL 101ST CAB

Densely populated areas of Mosul, Iraq.

Serving in OIF with the CAB was a dream. They were lethal on the battlefield. I can never stop repeating that they had an incredible, robust mission, providing aviation support by flying soldiers assigned or attached to the 101st Airborne Division in Iraq. They also provided support in Afghanistan. Their helicopters hauled heavy equipment like tank engines. They were crucial in OEF and OIF, and nothing would have been able to move from one FOB to another without these heroes and their amazing AH-64 Apaches, UH-60 Black Hawks, Kiowa Warriors, AH-1W Super Cobras, CH-47 Chinooks, and several other aircrafts. I worked as part of the Adjutant General Corps family, but I learned to think like an aviator while assigned to the 101st CAB. They were smart warriors. Their diligent and hardworking aviation technicians working on the helicopters every day at COB Speicher were beyond compare. I could not think of any better way to reward their dedication to service than to efficiently and effectively process their requests regarding financial issues and ensure that they got paid on time, received emergency leave, and had a rapid turnaround on other pertinent personnel actions. They were also known to be generous at heart. Some aviators donated their whole, hard-earned bonuses to the Combined Federal Campaign (CFC), the only charitable campaign allowed in the federal government. I know this, because our office ran the campaign during OIF. CFC is the world's largest and most successful annual workplace charity campaign, with almost 200 CFC campaigns throughout the United States and overseas raising millions of dollars each year. The pledges made by military members, federal civilians, and postal donors during campaign season support eligible non-profit organizations that provide health and human service benefits throughout the world, such as St. Jude Children's Research Hospital, warrior projects, homelessness programs, food banks, SPCAs, and thousands of other charities. The director of the office of personnel management normally designates responsibility for the day-to-day management of the program to its CFC office. Three years after I came back from Iraq, I found myself as an active participant in promoting the CFC.

OIF-2 R&R

In May 2006, I took R&R in the United States after SSG Warren returned from R&R visiting his family in Tennessee. He mentioned that he was house hunting and told the S-1 team about the particular house that he liked in Clarksville, Tennessee; however, there was a depression at the back of the house, so he did not buy it. He showed a beautiful picture of the house, and I began to do research to obtain further details. It never occurred to me that I would be buying a house online. The agent did a wonderful job in the advertisement, and it was such an attractive deal. In short, I called the agent and inquired about the possibility of viewing the house further and possibly buying it online. The name of the agent was Sheryl, and she was a Filipino American who was then married to a military man and had two wonderful children. We remain friends to this day. I finally took R&R and met Sheryl in person. She was such an authentic person and an energetic real estate agent. I said I was homeless and looking for a place to live. She thought it was a joke, and she smiled. She showed me the house, and except for the sort of depression in the backyard, the house was appealing; it was clean and had originally been owned by a couple with an ethnic German background. Now I understood why SSG Warren did not buy the house. He mentioned that the depression would be dangerous for his small children, but that was solved relatively quickly when the city started planting trees over the depression. I also put big rocks on both edges of a small canal situated about 20 meters from one side of the house to prevent further erosion. The neighborhood was fine. I was told that many enlisted soldiers and officers lived in that area. The house was located 10 minutes from Fort Campbell and five minutes from the Governor's Square Mall on Wilma Rudolph Boulevard in Clarksville, Tennessee. Later on, an elementary school was built nearby. Although I did not have small children, I still had to consider the high school and other valuable information for future reference. I told Sheryl that I would think about it a little bit more, but I ended up buying the house. After returning to COB Speicher, I informed SSG Warren that I had seen the house he had mentioned and confirmed that it was clean and beautiful and that I had ended up buying it. He became excited and loudly extended his congratulations to me, suggesting we should have a celebration after redeploying to Fort Campbell. I agreed, and so did the whole team.

Christmas at COB Speicher

I was a lector for the Catholic services, and I coordinated readings for the Christmas midnight Mass and vigil. Our pastor gave a remarkable homily. He preached that it is all about believing in God and loving your neighbor as you love yourself. He also preached that Jesus is the reason for the season. The live Nativity was grand, with soldiers participating to commemorate the birth of Jesus. The readings, the caroling, and the Nativity were all joyful and an occasion to remember for a lifetime. This might be the only Christmas we would have in the battlefield, so we put our best foot forward and happily celebrated Christmas at COB Speicher like one family, away from home. The dining facility was also decked with Christmas decorations, and the chefs cooked steak and lobster. The brigade commander and

PART II: Second Deployment. Destination: Tikrit, Iraq

aviation officers served lunch in the line. This is a tradition in the military that is being followed to this day.

My team and I went to the dining facility together for lunch, and our S-1 boss was such a lively officer to watch. She was spiritually inclined and had such a great sense of humor. There was never a dull moment with her. She was always in meetings and working late hours, so the fact that she was spending her time with us for dinner was quite an honor. On the other hand, our lieutenant was always cheerful and smart. She would tell someone right away if something was wrong or if it was exactly right. She was fierce, and it would not have been a surprise if she had earned her two bars soon. She also had a good sense of humor. Our chief warrant officer was also lively. It was hard to tell if he was joking or serious because he was always smiling. But he was a bright warrant officer. SFC Schroeder and the chief exhibited boundless excellence. One day, a brigadier general commented that SFC Schroeder had the potential to be a warrant officer. He might as well have been, being a sharp learner and a doer. SSG Warren and the lieutenant had something in common. They could debate each other on issues but still end up being friends. SSG C was cool, and we could all trust that, on her watch, everyone would be on their best behavior and there would be no issues. She could have become a drill sergeant if she wanted to. I could see a drill sergeant in SGT Broadnak too. I did not want to make him mad, although sometimes, he proved to be a good comedian. SPC P was a responsible specialist. She knew the drill. Private First Class (PFC) C was a live wire and was always cheerful. She was always a positive thinker and got along well with others. PFC D was the youngest member of our team, and she was sweet and smart, but we always protected her from the men because of her vulnerability.

I was the NCO in charge of this team, and I felt proud as we all sat for Christmas dinner and exchanged hilarious jokes. Dr. Diaz, our 101st CAB physician, approached our table and made constructive comments. He said that we were the best team in town and that we deserved the best dining table because we worked the hardest in the aviation team. Our S-1 said, "Of course!" Then we all laughed. Dr. Diaz and our S-1 boss were close, and they were both the rank of major at the time of our deployment at COB Speicher. We wished that they would marry, and our S-1 agreed, but of course, she was just kidding and went along with our joke. We were grateful for this occasion, a home away from home. In silent freedom, I prayed to God that He would continue to keep us safe and healthy and also protect our loved ones and supporters at home. The camaraderie in Iraq later extended beyond redeployment, and we were reunited at our S-1's house for a barbeque. After finishing the brunch at DFAC, we all saw the media outside the dining facility recording a video of soldiers who wanted to send Christmas greetings on TV. I was tempted to do it too so I could send greetings to my family, but there was a long line. I planned to call my family anyway, so I left the queue and gave my spot to the next soldier in line.

USO

I recall Christmas 2003 when we were in Mosul. The holidays were a popular time for USO (United Service Organizations, Inc.) tours to visit the war zone during OIF. The USO is an American non-profit charitable corporation that provides live entertainment, such as comedians, movie actors, and musicians; social facilities; and other programs to military members and their families. We heard

PART II: Second Deployment. Destination: Tikrit, Iraq

about the USO tour in Camp Victory, Baghdad, Iraq, with Stephen Colbert, Charlie Daniels, National Football League players, and World Wrestling Entertainment superstars who performed an excellent show for the coalition troops during OIF-1. We missed that event; however, we were visited by popular celebrities such as Bruce Willis and his band at the division main and rear command posts in Mosul. I vividly remembered the photo-op with Bruce, and I wished I had a video recorder. How many opportunities are there to have a photo-op with famous Hollywood entertainers like Bruce on the battlefield? But we were at war, and I did not have a camcorder at the time.

While deployed at COB Speicher, we were visited by celebrities such as Kathy Griffin; her husband, who is a U.S. Marine Veteran; and their friends. Once again, I had a picture taken with Kathy Griffin, and in the background of the picture were the S-1 and her lieutenant. My soldiers and I met Kathy Griffin and company while they unloaded from the military aircraft at COB Speicher Airfield. The excitement spread like a wildfire throughout the 101st CAB, so everyone started to come over to meet them. Kathy and company presented a show that night in a park near the HQs that was attended by many soldiers and contractors. They provided energetic and witty entertainment, including both clean and dirty jokes. We mostly enjoyed the dirty jokes by Kathy Griffin. The show concluded with T-shirts and hats tossed into the crowd by Kathy and company. Everybody at the event enjoyed the show thoroughly and continued to talk about it in the following days after the entourage had departed Iraq. It was indeed a memorable evening, and I thanked the Lord that there were no enemy attacks on that night.

FRIES Training

WINGS OF DESTINY MAGAZINE * 101ST AVIATION BRIGADE *

Sliding past the sun. A Pathfinder is silhouetted in the desert sun as he fast ropes out of a UH-60 Blackhawk helicopter from 5th Battalion 101st Combat Aviation Brigade during FRIES training June 19 at Contingency Operating Base Speicher, Iraq.

Continuous training is a necessity in Iraq. When not conducting a mission, the aviators continued to train to enhance their skills and proficiency. The picture above shows a Pathfinder in fast rope insertion and extraction system (FRIES) training. For the Pathfinder Company, the only infantry company attached to the 101st CAB, and the pilots of 5th Battalion, 101st CAB, fast roping is not a Hollywood stunt, but a technique they may have to employ on a combat mission. FRIES would be utilized in a situation that called for soldiers to get into a tight spot quickly where an aircraft is unable to land. FRIES is one of the essential skills that Pathfinders must keep current and conduct training on at least every six months.

Rocket Attack

On March 16, 2006, an RPG hit two soldiers at COB Speicher. SPC Pinson and PFC Gonzales, both from the G-2 section, lost their lives as a consequence of the impact. Three years later, we honored these special soldiers with the military intelligence soldiers serving at the site. The passing of SPC Pinson and PFC Gonzales from the 101st Airborne Division weighed heavily on the emotions of all who knew them and became personal for the 101st Division. The duty of bringing in the man responsible for the attack was later passed on to the 25th Infantry Division, which assumed command of COB Speicher when the 101st redeployed to its home base at Fort Campbell, Kentucky.

More Casualties

Aircraft moved stealthily to avoid being blown up by RPGs. However, there were also aircraft casualties from mechanical failures, bad weather, and shootdowns. One of these incidents involved five crewmen and two pilots. Their aircraft went down in bad weather, and no one survived. We mourned the loss of these courageous warriors and hung their pictures on the wall to memorialize their service to the military. Another loss was that of a captain who died for a non-combat reason. It was horrible and shocking when we heard about it. I remember seeing the captain in one of the services at Destiny Chapel. His loss was depressing and was the talk on base the whole week. We held a memorial service for him. It was sad, but nonetheless, we had to continue with the mission and kept marching on.

Easter Sunday

On the eve of April 16, 2006, we celebrated the Paschal Mass of the Sunday of Resurrection, or Easter Sunday. "Christ has overcome the power of death and shown us the way to eternal life." These were the powerful words spoken by our priest during the Easter Sunday vigil at Destiny Chapel, COB Speicher. According to most ancient tradition, this is the night of keeping vigil for the Lord (Exodus 12:42), when, following the Gospel admonition (Luke 12:35–37), the faithful, carrying lighted lamps in their hands, should be like those looking for the Lord when He returns, so that, at His coming, He may find them awake and have them sit at His table. The liturgy of the Easter Vigil is tremendously rich and normally includes baptismal rites for the catechumens; the blessing of the new fire; the blessing of the water with which the catechumens will be baptized; and the Exsultet, or Easter Proclamation. Our priest explained that the prayer for the blessing of the fire reminds us that, through Jesus Christ, "God bestowed upon the faithful the fire of His glory," expressing the hope that, through our celebration of Easter, we may be so inflamed with heavenly desires that, with minds made pure, we may attain festivities of unending magnificence.

There was also a blessing of the water that is used for baptisms after Easter Vigil. The rite of blessing said over water by the priest to make it holy contains prayers of exorcism. It can banish demons, heal the sick, and send unwarranted grace upon all, yet, most of the time, individuals make the sign of the cross with this water without even thinking how holy it really is. Holy water can be used to bless people, places, and things that are used by the people in their goal of glorifying God with their lives.

The priest further explained the meaning of the Exsultet, which calls the earth to be glad as glory floods her, ablaze with light from her eternal King. However, we did not have the Exsultet at church. Also, we were given general absolution from a few traditions, such as fasting, due to the fact that we were deployed in a combat theater.

We prayed that the love of God may so enflame our hearts that we can share His love with the world, even while in a theater of conflict such as Iraq and Afghanistan, and that we can live with such zeal and dedication that we can someday share in the gift of eternal communion with God, the "festivities of unending splendor" that the saints enjoy in Heaven. During Lent, the focus is on the Kingdom coming, not on the Kingdom already having come. To emphasize the penitential nature of the spiritual journey toward the Second Coming of Christ and the future life in Heaven, during Lent, the Catholic Church removes the Alleluia from the Mass. It is normally sung before the reading of the Gospel, or the Good News. The Alleluia comes back triumphantly on Easter Sunday, or rather, at the Easter Vigil on Holy Saturday night, when the priest chants a triple Alleluia before reading the Gospel, and all of the people at Mass respond with a triple Alleluia. The Lord is risen, the Kingdom has come, our joy is complete, and in concert with the angels and saints, we greet the risen Lord with shouts of "Alleluia!" The picture below was taken during the observation of Easter Vigil at COB Speicher.

Easter Vigil, 2006, at COB Speicher. Blessing of the fire and blessing of water.

THE EASTER CELEBRATION AT THE AIRFIELD

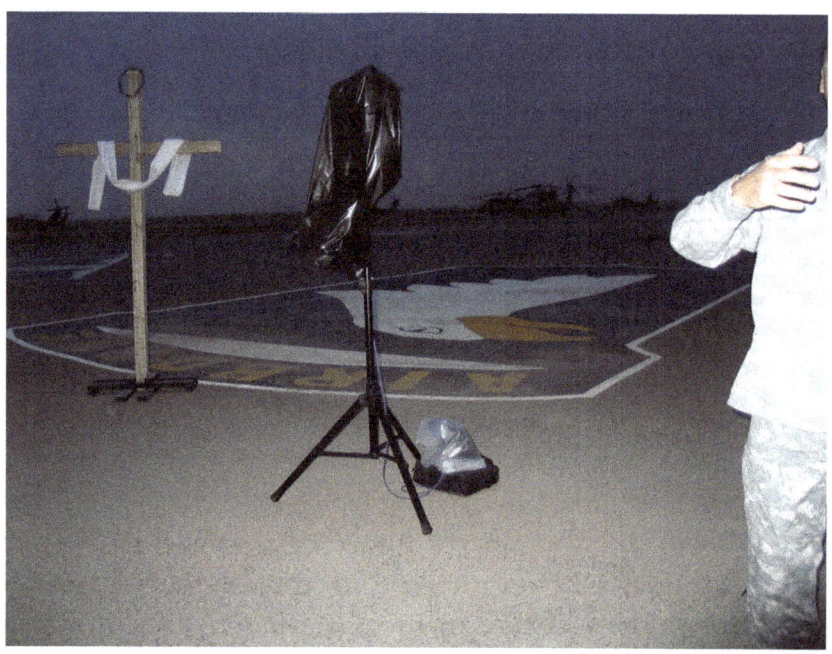

An early morning Easter celebration at the airfield, COB Speicher.

At sunrise on April 16, 2006, a non-denominational service was held at COB Speicher Airfield that was attended by many soldiers. Major Harrington read the scripture in the Bible, and Major Weicher, the pastor and celebrant, exclaimed that "Christ is Risen!" to which we replied, "He is Risen indeed!" Major Weicher further explained the symbol of the cross and the white cloth draping the cross. The placement of a white cloth on the cross on Easter Sunday represents purity and wholeness through Jesus Christ's resurrection. The cross is probably the most powerful and recognizable symbol of Christianity because of its significance in terms of Jesus's crucifixion. Christians see it as a symbol of Christ's victory over death. I remember the story from the Bible study led by Major Weicher: Mary Magdalene was the first to discover that Jesus's body was gone. She had gone to the tomb with the simple desire to anoint his body and shed a few tears of sadness and loss. But she was astonished when she saw that the stone had been removed from the entrance. She then ran to Simon Peter and the disciple whom Jesus loved, John, and said, "They have taken the Lord out of the tomb, and we do not know where they have laid him." Peter and the other disciple started for the tomb. Both were running, but the other disciple outran Peter and reached the tomb first. I remember the pastor's description of Peter's "sophisticated walk" instead of calling him "slow" when we discussed the reason that John reached the tomb first and Peter was running behind. Peter was probably walking carefully because of plentiful rocks on the ground. In those days, I could visualize a rough path that would hurt the feet a great deal.

John bent over and looked into the tomb at the strips of linen lying there, but he did not go in. Simon Peter came along behind him and went straight into the tomb, where he saw the strips of linen, as well as the cloth that had been wrapped around Jesus's head. The cloth was still lying in its place, separate from the linen. Finally, John, who had reached the tomb first, also went inside. He saw and believed. They still did not understand from scripture that Jesus had to rise from the dead. Then the disciples went back to where they were staying. Peter found some bread and told his fellow disciples to gather around the table in the same way they had gathered around Jesus Christ for the last supper. Peter broke the bread and gave it to the disciples. As they reenacted the last supper, Jesus entered and said, "Peace be with you." And everyone who was present believed. The exception was Thomas, who was not there during Jesus's first appearance to His disciples but who later believed when Jesus appeared again and let Thomas feel His hands and His feet. Thomas said, "My Lord and my God!" Jesus told Thomas, "Have you believed because you have seen me? Blessed are those who have not seen and yet have come to believe."

He is Risen; Easter Sunday.

AN UNFORGETTABLE EVENT

The story of the Resurrection is my favorite story in the Bible, next to the Nativity. I have been a lector in the Catholic church reading the scriptures since I was 18 years old, back in the Philippines, until this day at Our Lady of Angels, my parish in the United States, and at Saint Patrick's, which is about three blocks from my workplace in Washington, D.C. Reading scripture has been a burning desire of mine that can never be extinguished because the Word of God is the Truth that makes us truly free. The celebration of Jesus Christ's resurrection has affirmed and reaffirmed the most important bedrock of the Christian faith: Jesus's resurrection is the triumph of Jesus over sin and death. This is the triumph for all humankind who have been saved by the merciful providence of a good and gracious God who sent His only Son for just this purpose, the Son who fulfilled His mission on the cross and is now risen from the dead. The celebration of Jesus's resurrection at the airfield was very solemn, and we felt the power of the Holy Spirit among us as we offered peace to each other. It was a remembrance of a lifetime, a renewal of our faith as we battled through challenges in the theater of conflict.

PT AND AEROBICS AFTER LENT

After the Lenten season, I decided to join an aerobics class, also known as cardio, in the evenings after returning from work. I put on exercise clothes and running shoes and carried my little mat issued by the Army to the gym, which was located 50 meters from the LSA, for aerobics. The jacket came off after the warm-up and as the aerobic exercise became intense. At times, no equipment was involved. For one minute, we did lunges, push-ups—of course, dips, and torso twists, and then we started running in place, motivated by aerobics and workout songs. Dancing and kickboxing were so much fun; they were the favorites. The aerobics instructor made us sweat profusely and take advantage of the cardio program to improve our fitness. The DOD sent us a strong aerobics instructor from the United States, and she was a wonderful motivator. She dropped us down, brought us up, had us jog in place, took it to the left and to the right, and had us resume jogging in place. My understanding is that aerobic exercise stimulates the heart rate and breathing rate in a way that can be sustained for the exercise session. Aerobics has known benefits for both physical and emotional health. Aerobics exercise can help prevent or reduce the chance of developing some cancers, diabetes, depression, cardiovascular disease, and even osteoporosis. During the days when I was inactive because of my painful knees, I found that walking gradually was the beginning of a healthy life. I started by walking one mile a day, then two miles, and then four miles, and I remember walking 27 miles in a road march! The walks turned into running and led to a healthy mind and heart. Cardio helped me maintain my sanity while deployed in the middle of the desert in Iraq, like when I was at COB Speicher. It transformed me physically and mentally and from mediocre to a super performer at work.

PREPARING FOR THE PT TEST

Joining the aerobics class was voluntary, and the gym was always packed every time I went. I decided to give it up after a month because work started to pick up and I felt exhausted at night. When the temperature was not high, I continued to do PT in the morning, including push-ups, sit-ups, jumping jacks, killer curls, and a 2-mile run, and I started to prepare for the PT test. We conducted PT at 0600 hours, normally before the sun rose and before it got too hot. After warming up, we went for a 2-mile run near our LSA. By the time the PT test was a week away, I had to watch my diet too, in addition to getting ready physically.

Since basic training, I had always maxed my performance on the PT test. I remember my drill sergeant cheering for me on my last lap, and those cheers were such a great motivation. I ended up receiving a trophy for achieving the highest score on the PT test. I owed it to Drill SGT McCroskey for pushing me harder until I got mad and perfected the exercise, just so he would leave me alone. I recall the time during basic training when I started with one bad push-up. Drill SGT McCroskey kept counting one, one, and one. He never went past one. During every morning formation, he had his trainees drop to do push-ups until they got it right. I remember practicing push-ups while I was waiting for my orders for basic training. But they were bad ones. First, I was not maintaining proper body alignment. Sometimes, I would drop my hips to the floor; other times, I would drive my hips into the air as if I was pressing into a downward dog pose. I was told that I had broken my biomechanical chain connecting my upper body and lower body. What made it worse were my worms and head bobs, which made executing the up-and-down motion of a push-up look more like bad dance moves than a real exercise. My second mistake was shrugging my shoulders and having very wide elbows. My hands were way outside my shoulders and tried to initiate my movement with a wide pushing motion. It was painful, and I struggled to push up. I later learned that wide hands and bowed elbows forced me to shrug my shoulders and contorted my elbows. It was a weaker variation and caused shoulder pain over time. So, I tried to perform a standard push-up by aligning the webbing of my thumbs with my shoulders. As I lowered my body, I swept my elbows to my sides. I also kept as much distance between my ears and my shoulders as possible. Over time, I thought that I had improved my push-ups, as I was able to transfer much more strength through this motion. My third mistake was not moving through the full range of motion. My recruiter told me not to cheat myself when doing push-ups. It might be a bad push-up, but try to do it right and the next push-up will be better. There are two kinds of cheaters: bottom-half and top-half. I remember that my recruiter kept telling me that, if I were a top-half cheater, I would become weaker and would not have the strength to recover after reaching the bottom portion of a repetition. Staying high in the push-up movement typically means maintaining an arm angle between 180° and 135°. If I were a bottom-half cheater, I would likely be strong enough to do a quality push-up, but probably not as strong as I thought I could be. For this reason, I would end up bouncing between an arm angle of 90° and 135°. Sony, my late husband, was working in the Marshall Islands when I signed up for the Army. Nobody else trained me, and my fate in PT was left to the kind and gentle watch of my two recruiters from the Hawaii Kai Army Recruiting Station. They wanted me to be tough and patiently coached me until I reported to basic training.

Drill Sergeant McCroskey

Unfortunately, it turned out that my push-ups were not good enough for Drill SGT McCroskey. But I finally got it right and went up for the challenge. Drill SGT McCroskey had to watch closely as I lowered myself down to do a push-up and then rose to do more push-up repetitions. He liked yelling, and he would say, "Not good enough soldier. Lower!" And if I accomplished one push-up according to his standard, he would say, "Keep working on it, soldier!" He was proud of the fruits of his labor, as he continued to yell and scream at us day and night before we went to our bunk beds for lights out. I was lucky if I got some good rest, as I expected more screaming and yelling the next morning. I wondered if he was taking a lot of voice lessons to keep that commanding voice all the time. Drill SGT McCroskey started before dawn, and he was with us until it was time for lights out. I closed my eyes tight and wanted to get some sleep before the lights came up again, and I heard some more yelling and screaming from the drill sergeant. He was the only drill sergeant who was a screamer. The others were strict disciplinarians but not screamers. The one who really scared me was the senior drill sergeant. He was African American, and I stood still whenever he looked at all of us. He did not need to speak. His facial and body expressions were enough to bring us to a standstill. He had a smoldering voice that led us to strictly obey his drill sergeants, for fear that we would be kicked out and sent home. I think most of my fellow trainees wanted to go home. But it was just a threat, and in the end, we all successfully completed basic training. We never wanted to mess with the senior drill sergeant. Never!

Silent Freedom at Its Best

As trainees, we could not say anything against Drill SGT McCroskey. We developed our silent freedom in basic training; at least I did. I was silently angry about his screaming and bullying, but I kept it to myself. Drill SGT McCroskey was supposed to always be right and perfect, and we could not say anything to suggest otherwise. I remember the day when we had just finished training and I wanted to go the ladies room. I asked for permission, and he screamed, "What soldier?" I repeated that I wanted to use the bathroom. He said, "There is no bathroom, and what would you do in the bathroom?" I then remembered that I was supposed to use the word "latrine." I corrected myself and said I wanted to use the latrine. Then he let me go. I could not even roll my eyes at a drill sergeant, as it would cost me a lot of push-ups or sit-ups. As I finished using the latrine and coming down the stairs, a drill sergeant was heading my way. It was not Drill SGT McCroskey; it was somebody else from another platoon. I stopped and assumed parade rest. I heard the drill sergeant say, "Carry on," so I resumed running toward my platoon. I was not aware that they had been watching me when the drill sergeant was heading my way. They were laughing and told me later that they had been wondering what I would do. Some of them had secretly made a bet that I would ignore the drill sergeant and keep running down the stairs. That would have been a bad move and might have cost me an infinite number of push-ups. Well, it turned out that the other soldiers might have won. It was actually scary to see oncoming traffic made up of drill sergeants, so I was better off staying in the corner where no drill sergeant would be expected to pass. But the barracks was a bad place to stay when on break, because drill sergeants were all over the building. It was just like the cadence: "Everywhere I go, there's a drill sergeant there." So, I just kept my cool and maintained my silent freedom.

I lost a significant amount of weight during basic training. I went from 120 lbs. to 90 lbs. I was eating only rice and apples, though I did drink water and pure black coffee. I have read and heard that black coffee without sugar is good for the heart and can prevent diabetes. The taste was bitter, but I got used to it. I joined the U.S. Army four months after arriving in Honolulu, Hawaii. The food in Hawaii was similar to Filipino food, so I ate healthily while in Honolulu. It took me about two months in basic training at Fort Jackson, South Carolina, before I got used to the other food in the mess hall, and I started to drink less coffee, especially after singing cadences that said it tasted like turpentine. I started to drink milk and carbonated water instead. The food in the mess hall was supposed to get better the longer I stayed in the Army.

Drill Sergeant's Joker Grin

I maxed my performance on the PT test after that, and before graduation, Drill SGT McCroskey said that he knew I could do it and that he knew I was preparing for Officer

PART II: Second Deployment. Destination: Tikrit, Iraq

Candidate School (OCS). Drill SGT McCroskey would tell us, "Don't forget my name," and then he started spelling his name while we were doing push-ups and climbing up and down the steps in the barracks early in the morning and again in the evening. Nowadays, I just laugh it off whenever I remember those bad days. It's all about attitude. Whatever the drill sergeant says, do it or you will be on his bad list. I tried to maintain a good attitude and was never on his bad list.

I have lots of memories from my basic training, and I can even do a comedy skit about it nowadays. I actually enjoyed it. It gave me self-confidence and sportsmanship, taking almost nothing personally from Drill SGT McCroskey's yelling, screaming, and talking down to his trainees. He really made us feel like dirt, and when we successfully graduated from basic training, he was the first one to congratulate us and say, "Good job!" I was used to seeing him with a mean face, so his grin came as a surprise, and he almost looked like the Joker. He said that he was so proud of me for maxing out on push-ups, sit-ups, and the 2-mile run on the PT test and emerging as the top scorer. I had something to be proud of and would never be afraid to be in a PT contest. I did not smile when he was talking from under the brim of his drill sergeant hat, and even though I had already graduated from basic training, my discipline still held. I always answered, "Yes, Drill Sergeant," when he was talking to me. He shook my cold, sweaty hands after he talked and wished me good luck on my advanced training. I said, "Hooah, Drill Sergeant!" He looked at me one more time and then turned away. I finally relaxed and hoped that I would never cross his path again. Not within the Fort Jackson perimeter anyway. Not in the PX, not in the commissary, not downtown, not anywhere. I wanted to forget his grin.

Asian Pacific Month

Asian Pacific Month was celebrated by the 101st CAB on May 26, 2006. Many talents came out on that day in celebration of Asian Pacific heritage. SGT Siler from HHC Company, 1st Battalion, 101st CAB, performed a Samoan dance as part of the 101st CAB's Asian Pacific Month Program held at Destiny Chapel at COB Speicher, and her Samoan admirers adored her during her Samoan Diva dance. Chief Warrant Officer 2 Weyrauch, one of the aviators from Company C, 5th Battalion, 101st Aviation Regiment, also performed a ceremonial dance during the program. The celebration was a huge success. The dining facility had Asian Pacific food on the menus for a full week. I would have cooked the favorite of all, lumpia Shanghai, if I had been allowed to. Lumpia Shanghai is a Filipino deep-fried treat consisting of a mixture of ground pork, carrots, and salt and pepper seasoning wrapped in a thin egg crepe. They are also popularly known as egg rolls and make excellent appetizers.

The celebration reminded me of the Asian Pacific Month celebration that I coordinated at Fort McCoy, Wisconsin, so many years ago. I wanted it to be memorable, and I wanted its impact to last for a long, long time, especially for the civilian community, since we, the military, rotate duty assignments and never stay in the same place for very long, typically not more than two or three years. The Asian Pacific Month event that I organized was held at the local community center, which provided a large enough space for many attendees. There was a potluck and displays of various crafts that were made in Asia. I remember preparing pancit (noodles), the renowned lumpia Shanghai, and chicken adobo. Chicken adobo has a glaze that is savory and sweet with a hint of tang and a distinct soy flavor. Garlic and onion create a savory base along with bay leaves, and peppercorns add subtle pops of heat. All of the food I cooked was a hit. I could have roasted a pig too, and maybe they would not have allowed me to leave because I started a feast. The food tasting was followed by a hula dance contest. The participants were wearing Hawaiian wraps, grass skirts, and other traditional accessories. One of the male participants wore a coconut shell bikini top and raffia skirt, which added so much fun to the event. I was wearing a Hawaiian wrap with a flower in my hair and a garland. The music played, and the contest began. There were winners and losers, but everyone appreciated the camaraderie. There was also a raffle drawing with prizes in the form of gift cards to the PX, Walmart, and other stores. These gifts were donated by the stores. The event concluded with hula dance participation from the audience. It was a family gathering, so the children were dressed in Hawaiian grass skirts for the girls and Hawaiian shirts for the boys. All of the comments were urging that the celebration be made an annual affair. It was an occasion that people will remember for a long, long time.

Hellcat Farewell BBQ Bash

A banner prepared by 101st CAB HQ for the Hellcats.

The soldiers from HHC, 101st CAB, decided to ask their commander for a farewell party before redeployment, to be held on the Fourth of July. The commander approved, and everyone just had a wonderful time, though still remaining vigilant of the dangers at COB Speicher, Iraq. The party was to support troop morale and to congratulate the troops for a job well done. Hamburgers and hot dogs were the favorites of the day, and the first sergeant and company commander served. An ice cooler full of near beers, sodas, and bottled water was also available. It was a hot, muggy day, so we decided to celebrate the occasion indoors and put camo nets outside as protection from sunburn for the sun lovers.

Preparation for Redeployment

The month of August finally came, and everyone was excited as they started preparing for the redeployment. We received security briefings, which were similar to the briefings from the last year. For the people who had already had multiple deployments, the preparation was routine. They knew what would happen next after the briefings, such as mailing home excess baggage and carrying only what was necessary. It was easier for the leadership when the soldiers already knew what to do to prepare for redeployment. At least we did not have to worry about a ground convoy this time, so it was not as stressful as the previous deployment. Luckily, some people did not have excess baggage, so they did not have to worry about shipping anything. I made sure that everything fit in my two duffle bags and backpack. Additionally, I had an M4, which was easier to carry than the M16A2. The M16 was almost as tall as me, and I had to readjust the length so it would not touch the ground. This time, it was light traveling for me.

We were informed that the 25th CAB would replace the 101st CAB at COB Speicher. We were excited, as we would be free from the bad guys, the sand, the dust, the creeping

PART II: Second Deployment. Destination: Tikrit, Iraq

insects, and especially from the depression. We welcomed the idea of flying as well—no ground convoy.

Soldiers of the 101st Airborne Division's Combat Aviation Brigade and the 25th Infantry Division's Combat Aviation Brigade display the colors during a Relief in Place ceremony at COB Speicher, Iraq.

CHANGE OF COMMAND CEREMONY

A Change of Command ceremony took place in Tikrit, Iraq, on August 30, 2006, to transfer responsibility for air operations in northern Iraq from the 101st Airborne Division to the 25th Infantry Division. I was familiar with the 25th CAB, which hailed from Wheeler Army Airfield, Honolulu, Hawaii. On that day, they relieved the 101st CAB, which had served at COB Speicher since August 2005 in support of Operation Iraqi Freedom. Like the 101st CAB, the 25th CAB was no stranger to deployment. Specifically, it served in Bosnia in 2002 in support of Operation Joint Forge, and it also served in support of the GWOT in Afghanistan in 2004 as part of OEF and OIF. Elements of the brigade were also deployed to support humanitarian efforts for earthquake victims in Pakistan in 2005. As part of the U.S. Army's transformation, the brigade was renamed 25th CAB after its reconfiguration and acquisition of two new battalions in 2005. The 25th CAB's mission was to provide air support to the 25th Infantry Division units, which had the larger mission of working by, through, and with the Iraqi security forces in the interest of a safer and more secure country of Iraq. The event also marked the completion of the 101st CAB's second deployment in three years. The 101st CAB supported the GWOT in Afghanistan in support of OEF/OIF from August 2003 to August 2004. In 2005, the 101st CAB accumulated more than 110,000 flight hours by conducting recon missions and air assault missions supporting both Iraqi and U.S. forces. Our brigade commander, Colonel Warren Phipps, remarked, "Not only am I proud of my soldiers, but I would like to acknowledge the growth in professionalism, discipline, and confidence in the Iraqi Army. They have made considerable progress in the last year." I felt proud of our brigade commander and soldiers too, especially for treating my team well, just as they took care of their aviation team. Truly, they are 101st's Wings of Destiny.

Redeployment First Sergeant

We redeployed to the United States in batches. While we were in Kuwait, the commander needed a first sergeant to help him with accountability. I stepped forward, told him he could consider it done, and took charge. There were more than 300 troops on the manifest, and half would be dropped off halfway to Fort Campbell. Together with my assistant, we accounted for the troops that would be redeploying based on the manifest. We conducted two formations: one before we put the soldiers on a short break and one after the break when it was time to board the buses. We conducted a roll call and found that two soldiers were missing in the second formation. I announced that no one would be going anywhere until the two troops showed up. We would not leave anyone behind. Some of the soldiers were already whining, of course, because they did not want to miss their flight. We kept calling their names until, finally, they showed up. One of the soldiers had been stuck in the bathroom, and the other was in line at the PX and had lost track of the time of the formation. Luckily, this was not the time for any punishments, and we went through the manifest and ensured that all soldiers were accounted for. We then boarded the buses, with the commander and my assistant being the last ones to board. The buses took us to Kuwait International Airport for a flight bound for the Freedom Land, the United States. As soon as the plane reached its cruising altitude in the air, some soldiers started relaxing: some watched movies, some were eating, and some took naps. I started to relax too and was looking forward to seeing my family again. The flight was 13 hours and 46 minutes long.

"Welcome home" to troops from their deployment.

Welcome Home

The welcome home ceremony was extremely emotional, and as for the first deployment, it was full of excitement. From the aircraft, we saw families and friends with huge "welcome home" signs waving small American flags as the aircraft halted. We could hear more excitement and laughter as we descended the passenger stairs to soon be reunited with our families. The troops grounded their backpacks and then formed a formation for a roll call before marching into a building made especially for departure and welcome home ceremonies. The families were seated on the bleachers, anxiously waiting to hug their heroes. The leaders gave a brief talk and then dismissed the soldiers for the day, so they could reunite with their loved ones. It was great to be back.

Time Off

We were given time off after all of the post-deployment activities; it was relaxation that we needed most now that we were away from austerities of war-torn Iraq. It took a while before most of us adapted to our lives in the United States. For example, we did not have to dig foxholes anymore, we had shower rooms inside our homes, we no longer had to carry M4s, there were no sandstorms, there was plenty of running water, we now owned refrigerators and microwaves, and we did not have to worry about enemies lurking in the corners of the streets. It was home, and we were free.

After the redeployment, some of our soldiers left the Army due to expired terms of service—that is, because they had reached the end of their enlistment commitment. Others left for a permanent change of station, and the rest remained

PART II: Second Deployment. Destination: Tikrit, Iraq

with the 101st CAB. I remained with the CAB for about four more months before I requested to move to the 101st Airborne Division Resources Management (RM), which is responsible for managing budget and funds for the division. This position did not have the excitement of deployment to a theater of conflict, but I enjoyed the numbers and working on the financial resources. Accounting and financial management were my first love in the military. This area gave me the knowledge and experience I was looking for and enough to be employed after the redeployment. After 23 years of honorable service, I decided to put in for my retirement, and I now live in Clarksville, Tennessee.

The author during OIF-2, COB Speicher, Iraq, 2005–2006.

PART III
BACK TO IRAQ. DESTINATION: THE ZOO

The Victory Arch, which became the hub of operations in the Green Zone of Baghdad, Iraq, and next home for my OIF III.

Third Tour to OIF

The military aircraft was halfway done unloading its passengers from Kuwait to Sather Air Base, Baghdad, Iraq, when someone yelled "Get down!" We were startled, and we all hit the ground. Some were still inside the aircraft, others were in the middle of the airfield, and others were near the T-walls. I was caught between the T-walls next to an oxygen tank. We were told to duck, and we stayed in that position for about 30 minutes, until we heard someone yell, "Clear!" Everyone got up, and I told myself silently, "Wow! That's quite a welcome, Baghdad!" All of the passengers then proceeded to a huge tent and then out to the cages to retrieve our luggage. My luggage was not in the first cage. So, I went to the second and then to the third. I was relieved when I finally found my duffel bag, and it looked like it was undamaged and locked. I put the strap on my shoulder and proceeded to the waiting area to find my ride. It did not look like a fancy airport with a crowd waving and shouting. It was a quiet welcome compared with the excitement at the airfield a few minutes earlier when we all ducked because of an incoming RPG. If my transportation pickup was a no-show, I was thinking about spending the night inside the huge tent. I had to think of worst-case scenarios in my strange new host country and what my alternatives would be in order to survive; after all, we were still at war. If a problem occurred, I would have to call the company project manager and maybe he could facilitate things so that I did not have to spend the night in the tent.

I was still tapping my passport with my right index finger when I saw two gentlemen waving a placard with my name, so I walked in their direction. After exchanging greetings, they led me to a civilian car and gave me a bottle of water. With an Arabic accent, Mr. Arshad said, "Welcome to Iraq!" I replied, "Shukraan lak; kayf halakum?" Mr. Arshad was quite pleased with my Arabic and laughed. I had learned my Arabic from Dr. Ali during my first tour. The Iraqis seemed to be fascinated by my looks and by how quickly I learned their language. I felt that my karma was kicking in again as I talked to Mr. Arshad. Mr. Arshad and the other man were hired by a U.S. company for this mission. Mr. Arshad later became like a father to me, bringing me jewelry and other gifts every time he visited his family in Kuwait. I still treasure those gifts to this day.

The other man barely smiled, looked mean, and stayed quiet until they dropped me off at my new temporary home at the Baghdad Zoo, about a 25-minute ride from the airfield. The prospect of living at the zoo was not appealing, as there was a story that Saddam Hussein had fed the animals with people he disliked. This is a horrible story, but it was my

PART III: Back to Iraq. Destination: The Zoo

destiny to be assigned to the zoo. We finally reached the gate, and two guards asked for our identification cards. Under the light, I could see that the two men were muscular, and based on their physical characteristics, I concluded that they were from Fiji or Samoa. However, Mr. Arshad said they were from Uganda. I almost kicked myself for guessing wrong. Mr. Arshad laughed and said not to worry, as nobody is perfect. I was trying to create a good impression, and instead, I messed up. I thought I was good at assessing ethnicity and personality from physical appearance, but I missed this time, and I was embarrassed in front of Mr. Arshad. My guess fell short because I have several friends who are from Fiji and Samoa and I am familiar with their distinguishing features; they are muscular. My Samoan friends can easily identify my ethnicity because of my eyes and my accent. I was wrong, and I told Mr. Arshad I must be tired, and he said it was a good excuse, and we both laughed. The guards asked a couple of questions and then said, "Welcome to the zoo, ma'am." They were the second ones who called me "ma'am." Mr. Arshad was the first.

There was no crowd to welcome us, and it was dark. The guards directed me to my quarters and said that there would be a formation in the morning at the building next to my billet. I noticed another car dropping off the new linguists in the next building. We waved to each other, and they said goodnight. I replied goodnight and told them to sleep well. The building next to mine was very fancy. I found out later that the buildings at the zoo had been occupied by the CIA during the invasion of Iraq in 2003.

The Baghdad Zoo originally opened in 1971. It covers 200 acres and is located in the Al Zawra'a Gardens area of Baghdad along with the Al Zawra'a Dream Park (an amusement park) and Zawra'a Tower. Before the U.S. invasion of Iraq in March

2003, the zoo housed about 650 animals. However, the zoo was destroyed during the Battle of Baghdad in early April 2003. Because of the invasion of the city, the staff and officials left the zoo, which then suffered from severe looting. The cages were torn open by thieves who released or took hundreds of animals and birds. The zoo staff said that most of the birds and game animals were taken for food, as pre-war food shortages in Baghdad were exacerbated by the invasion. The zoo was restored and reopened in July 2003. Included in its collection was a group of lions reared by Saddam Hussein's son Uday that had been removed from one of his former Baghdad palaces and taken to the capital's zoo, where the capable staff took good care of them. Such views of Baghdad during peacetime are fascinating, and it would have been a broadening experience to tour there. However, I thanked God for being with me and keeping me safe at all times. Returning to Baghdad, Iraq, was an experience of a lifetime.

After thanking Mr. Arshad and his companion for the ride, I went to bed, and as usual, my first night there was a sleepless night. I asked myself what I was doing in this place. Things happened like a whirlwind, and before realizing it, I was back in the desert. I did not even tell my parents or my son about going to Iraq, so I planned to do it in the morning when I could take a break from work. I knew they would be stunned. It had been just a few days earlier when we had talked on the phone in the United States, and I never mentioned the job in Iraq. I was afraid it would fall through, so I never mentioned it to them. No one in my family wanted me to go back to Iraq anyway, because I had promised that the second combat tour would be my last. Yet, there I was. I had broken my promise not to return to this hot, humid, scorpion-infested, hostile place.

PART III: Back to Iraq. Destination: The Zoo

Reminiscence of the Past

As I gazed at the ceiling, my mind traveled to the time when I retired from the U.S. Army. In May 2007, I had tendered my request to retire after 23 years of active duty service in the U.S. Army. I was on terminal leave until my actual retirement date of August 31, 2007. I had mixed emotions about my retirement. I wanted to do more as a soldier and deploy to Afghanistan, but my body gave up on me during that period. Everything was painful, and I could hardly move. A voice told me to take a rest, which led to my retirement. I ended up with a lot of medications and quit running for a couple of weeks so that I would at least be able to walk. I ended up purchasing a 2,237-square-foot house in Clarksville, Tennessee, with a good-sized backyard. It had an oversized garage, and the original owners did a marvelous job with it. The storage was a separate building located in the backyard. The first level had hardwood floors, a good-sized kitchen, and a small living room with a fireplace and TV. I placed a pub table next to the window and watched the passersby. There was a half bathroom on the first floor. On the second level were three bedrooms and a huge bonus room over the garage. In the bonus room, I placed an office desk next to the arched windows and a convertible daybed. A huge TV completed the bonus room and made it cozy. One thing I noticed about houses in the southern United States is that one can get more for their money with a relatively inexpensive investment property when compared with properties in Washington, D.C., and its metro areas. The cost of properties in the city is greater: about four times more than the cost of properties in Clarksville, Tennessee. My home in Clarksville was too big for only one person living in it, but my son decided to stay in Michigan so he could be near his friends. This property was the house that Staff

Sergeant (SSG) Warren had almost bought when he took his R&R during the second deployment in Iraq.

I made several improvements around the house, not to the structure itself but to the depression that SSG Warren had mentioned. The backyard is now safer for people to walk on and for children to play in. The depression has been covered with grass and trees and is not as bad as I thought it was. I also had some large rocks placed over the eroded soil caused by the rainfall. The original owners seemed to ignore the effects of water erosion from rainfall by breaking soil aggregates, which causes the detachment and displacement of the soil, either directly by the impact of raindrops or indirectly through bodies of water, such as the small canal cutting through my property. I did some research on how to prevent further erosion and found that rock toes, or low structures of rock placed along the edge of the bank, layered with live staking, plantings, and seeding, would be helpful. I called a service provider to install the rock toes and was very happy with the results. The rocks placed on the slope helped stop the erosion and eliminated some of my worries. SSG Warren would have been proud of the improvements that I had made around the house. He would have been proud of me buying the property. I was quite proud of my work, too. The house could have been perfect for his wife and small children because it had a huge backyard with trees, a nice concrete balcony, and railings going down to a concrete patio. It would have been a nice setting for an evening outdoor movie, using the white wall as a projection screen. I fell in love with the house.

I rented the beautiful house when I deployed to OIF for the third time, and it was still rented when I returned to the United States. I decided to stay in Washington, D.C., and hired a property manager to take care of my property in

Clarksville. A few years later, I received notification that my next-door neighbor had built a chicken coop that encroached on my lot in Clarksville. I had taken a real estate class in Nashville and recalled the discussion about encroachment. Encroachment occurs when a party who is not the property owner builds a fence or other structure over the property line or plants a tree with branches that hang over onto the adjoining property. My neighbor built a fence over his property line and encroached on my lot. I was perturbed by the news, and if I let it go and did nothing, the encroachment could restrict my ability to transfer the title to my property when I sold it. It did not seem right, so I decided to visit the property and talk to my neighbor before things got worse to avoid litigation as much as possible. I believe in diplomacy.

I discovered that my neighbor was a successful truck driver, spending most of his time on the highway away from home. I decided to leave a letter in his mailbox requesting to talk to him and letting him know that I flew back to Clarksville just to speak with him. It was a cordial letter. My one-week vacation in Clarksville was almost over, and he had not shown up, so I decided to call him with the phone number that his wife had given me. Fortunately, he answered his mobile phone, and we discussed the encroachment. I gave him the measurements of my lot, and he said that he would take care of it when he returned, explaining that his occupation had him on the road quite often. He sounded like a gentleman and a professional. I told him that I understood, as I was a Veteran who used to be away from home, too, because of lots of field training exercises at Fort Campbell. Surprisingly, he said that he loved the troops and supported them wholeheartedly. I felt the genuineness of his words and hoped that the encroachment issue would be resolved sooner than expected.

We both hung up, and I was satisfied, although I had yet to find out when he would move his fence. I made several phone calls to follow up, and sometimes, there was no answer, but he returned missed calls and apologized because he had been on the road at the time. I was patient, and I thanked God for my silent freedom and for helping me to be understanding. Being a truck driver is like being a soldier: always away from home. And once at home, it is always a treasure to spend quality time with family. My neighbor was in his early fifties, and I admired his energy and commitment to work. Again, he reiterated that he valued our Veterans and supported them fully. I was touched by his words and glad that he was my neighbor.

As told in the Gospel of Matthew, one of the Sadducees, an expert in the Law, said to Jesus, "Teacher, which is the greatest commandment in the Law?" Jesus answered him, "You shall love the Lord your God with all your heart and with all your soul and with all your mind. This is the first and great commandment. And the second is like it. 'You shall love your neighbor as yourself.' On these two commandments depend the whole Law and the Prophets." (Matthew 22:36-40). I find peace whenever I think of these two powerful commandments, and I always felt God's presence when I talked kindly to my neighbor with the hope that he would fulfill his promise to remove the encroachment on my property. One day, I received a text stating that he had corrected his mistake and removed the extended barrier. The chicken coop was now all sitting on his lot. I thanked him for it. I went to Clarksville for another visit and saw that he had indeed removed the barrier and moved the fence onto his lot. I know that he had to spend some money on it, but it was an error that had to be corrected. I wished that I could meet him and thank him personally, but he was on the road again. However, we still keep in touch to this day just to exchange greetings. I decided to call my neighbor and thank him for keeping his promise. He was such a gentleman, and he simply said that he values Veterans and respects them for their sacrifice and service to the nation. I am proud of my neighbor.

After Retirement from the Army

After a couple of weeks in retirement, I felt better and started looking for a job. I was also preparing to become a real estate agent. One day, as I was exiting the gym at Fort Campbell, I saw a fellow Veteran who used to be assigned to the 96th Battalion, a unit of the 101st CAB. After exchanging hellos, I asked what he had been doing, and he mentioned his company and the work they were doing. I became interested, and he invited me over to his shop to observe. I showed up one day and felt that I wanted to be part of their mission. Before I knew it, I was interviewed by the company's project manager, who asked when I could start. It was a good interview, and it felt right. I replied that I was available immediately and showed up the next day to work as a contractor supporting the Rakkassans in the rear while they were deployed in Iraq and Afghanistan. Not being part of the action in the field felt strange, and I have yet to learn to live with it.

My fellow workers were also Veterans, and we all got along well, having served in the same 101st Airborne Division. I was happy to be surrounded by a family of Veterans. As days passed, I started hearing about jobs overseas. Deep in my heart, I said silently, "I have been there, done that, got a brown T-shirt that says Al Asad, and will never go back." But my words contradicted my heart's desire. My silent freedom wanted to scream and go back to the battlefield. I wanted to get back to the theater of conflict and to serve again as long as the U.S. troops were there. I felt more secure knowing that I would be surrounded by fierce, capable troops. They have been trained to be lethal on the battlefield, and I have always felt that I was in good hands with them. After listening to the same conversation at my work every single day, I decided to check it out and ask what it was about the jobs overseas that was making the others so excited.

One of my co-workers showed me the website and how to access the list of jobs available at that time. I noticed that my colleagues had been dropping off their resignation papers one by one and successfully landing jobs overseas. Their gradual departure whetted my interest, so I became more curious and seriously searched for a job that I would like. Simultaneously, I continued preparing for the real estate agent's test and felt that I was ready for it. I planned to join my real estate friend Sheryl, and together, we would build a real estate company catering to Veterans and the community in Clarksville. It was a grand idea. But my curiosity distracted me from my ideas, and I continued to search the websites. I found a certain job opportunity that matched my experience in the Army, and I jotted it down. I remember sitting back, getting up from my chair, and beginning to walk and ponder the job in Iraq. The opportunity was compelling, and my silent freedom was again screaming to take the job. I strongly believed that I was healthy enough and could serve again. I made an argument silently, only to be defeated. I had to go.

While walking in my neighborhood, I thought about my accomplishments after returning from my second tour in Iraq supporting OIF. I had settled down in Clarksville and continued to finish my last courses online. My mother-in-law Jane and I continued to communicate even after my divorce from her son. She was a wonderful lady who supported me emotionally during difficult times. She loved and honored our Veterans. My father-in-law and I also continued our communication after the divorce. Both he and Elva, his current wife, also emotionally supported me and my son during the challenging times. My father-in-law Bill was the executor of my last will and testament in case something happened to me. My son was nine years old at the time. Bill was a U.S. Navy Veteran himself. Both he and Elva continued to support the troops after he left the service, and they fully loved and respected our Veterans. One of their sons-in-law, Russell, also retired from the U.S. Marine Corps, where he had served well. We are a family of Veterans. I first met Russell when he was still an active Marine and was assigned to Honolulu, Hawaii. I was assigned to Fort Shafter, Hawaii, at the time. He was with his family, and meeting them was a significant event for me. I had been assigned to Fort Shafter for four years and was waiting to join my former husband after he completed the resident course at the Warrant Officer Candidate School at Fort Rucker, Alabama.

Before I was done thinking about the past, I had completed a lap around my neighborhood and stood in front of my house. I asked myself how I could leave such a neat house. I decided to sit down in my foyer to drink a bottle of water. The neighborhood was quiet, and I noticed that the weather was just perfect. The man who lived across the street from my house was retired Special Forces, and I had seen him mowing his lawn. His name was Andre. We had waved hello to each other a couple of times and had a short, friendly conversation. He had a wonderful ranch house. Farther down the hill were more beautiful houses and good people. My house sat on top of the hill with red brick in the front, and I had built a mailbox resembling a mantle clock sitting on red brick, like the front of my house. It was unique, and the idea had originated from a birthday gift. The builder did a perfect job on the mailbox, and it was worth the investment. It gave me joy every time I looked at it. There were three tall, thin trees in the front of the house, and I had decorated them with Christmas lights in different colors during the last holiday. Some people stopped at my house to tell me how much they appreciated the colorful LED lights: warm white alongside shades of yellow, green, and red, perfect for Christmas. I was happy to give joy to my neighbors in even a simple way. I was thinking that maybe I should decorate the trees all year round with soft white lights and add to the bright stars up above. My ideas of decorating originated from an innate interest that I should have cultivated and that might have helped many business developments, such as adding to the real estate business. It sounds very interesting to me to this day. Oftentimes, my kids admire my creativity in keeping the house lively and in perfect shape for selling, a hobby I should have pursued.

Fruits of Labor

In July 2007, I decided to call the contact for the job in Iraq and ask whether the advertised job was still available. I intentionally sounded jovial, although I was not feeling confident because I had waited for a few days, so the job might no longer be available. To my surprise, though, the man on the phone said that it was still available. I was more determined this time and applied for it. I followed up with a phone call and asked if my application had been received. The man said yes and that he would send it to the management in Iraq for review. Three hours later, the fruits of my labor paid off when the same man called me back and informed me that the management liked my resume and had asked when I could be available to deploy. I was thrilled by the news, and without hesitation, I said, "Right away." The man laughed and said that they would conduct a background check, which would take about two weeks. After we hung up, I called Sheryl immediately and consulted her about my house. She was confident and promised that she would take care of it immediately, so that I could go ahead and prepare for my upcoming third deployment to OIF. We were both ecstatic.

JOB TRAINING

In August 2007, I departed Clarksville, Tennessee, for job training. I reported to my new company in Herndon, Virginia, and finally met the project manager. He was also happy to finally meet me. He further informed me about documents that required my signature, so he asked if I could meet him at the D.C. office at a time that was convenient for me. The next day, I took a taxi to meet him at their headquarters. He introduced me to other employees and proceeded to the HR office. After reading and understanding the contract, I signed it, and they handed me my copies. The project manager saw me to my taxi, and I returned to Herndon to continue with my training. I met the other civilian contractors, who would soon join the linguist team in the battlefield. It felt weird to be deploying without my uniform and without my M4, the weapon I was accustomed to during OIF. Duffle bags and boots were necessary for the travel. At least I was wearing good boots, which were a lot better than my old combat boots during the first deployment and were comfortable to wear. I had worn boots for 23 years in the U.S. Army, and the new boots I was wearing were not comparable to the old ones. The new boots were the most comfortable boots I had worn since joining the Army.

I packed cargo pants, planning to buy the rest of my clothes later in the stores at Camp Slayer, Iraq, selling 5.11 brand tactical clothes. I found such clothes appropriate to wear in places such as the battlefield in Iraq, which is hot and humid and has sand, sand, and more sand. After the familiarization training, the company transported us from Herndon, Virginia, to Fort Benning, Georgia, where we were supposed to complete the medical readiness processing. We were asked tough questions, and our medical records were screened. I received a phone call while waiting in line at one of the processing stations. It was a call from an interview panel member from the Rock Island Garrison Arsenal located in the Quad Cities region of Illinois. I had a friend who worked there who encouraged me to try and apply for a federal government job after my retirement from the Army, so I did. I had a feeling that the man on the phone would offer me a job. However, I mentioned that I was in the process of deploying to Iraq. He was surprised to hear that and was speechless for a few seconds. He finally spoke again and gave me his support by wishing me a safe trip and best of luck, which I was sure would be much needed. However, what would happen if my readiness processing fell through? I had just declined a job offer from the federal government. Again, I told myself that reverse thinking works, as I have proven to myself so many times. Without further ado, I raised my chin up and went back in line to complete the medical examination process. Praise God, I passed with flying colors. My silent freedom was at the helm.

Up, Up, and Away

In September 2007, I remember feeling my pulse point on my neck as I boarded the plane to Kuwait. I kept reassuring myself that everything would be just fine. I called Sheryl for the last time prior to boarding the plane. She said that the house had been rented after I departed for D.C. I thanked her, and we ended up best friends for life. The pilot

PART III: Back to Iraq. Destination: The Zoo

announced that the flight from Fort Benning to Kuwait was 14 hours and 52 minutes. I decided to relax my mind as soon as we were seated in the plane and feast my eyes on the scenic view until I could no longer see anything. I focused on what I would eat as a couple of flight attendants started their routes and offered us snacks and drinks. I waited until the food was served before I forced myself to take a nap. I used my earplugs, eye mask, and blanket. I saw that the other passengers had decided to watch movies and play games.

I must have snored a little bit and was glad that I caught it before my snoring got louder. I felt uncomfortable after sitting down for so many hours, so I got up to stretch my legs. The plane was a huge Boeing 747, and it was convenient for walking back and forth and around the seats. I remember getting up and walking around 10 times before we landed, and the other passengers did the same. My silent freedom told me to kiss the ground after the plane had unloaded. But I said no, because it was not my first time and I was looking forward to Iraq. This was my third tour. I thought I was getting good at it, having been deployed to this place multiple times. It became a familiar scene to be bused to Kuwait and wait to be manifested to Iraq, and once in Iraq, it would be another drill. I reassured myself in silence, a Veteran on the move.

The staging area in Kuwait had improved a lot. Each tent had air conditioning, a necessary element in the 150°F heat. The tents were of different sizes; some housed four people, whereas the larger ones housed more. After the briefing, we were assigned to get our things together and start moving to the tents. By the time I got to my tent, three individuals had evidently picked the bottom bunks. I ended up having a top bunk. Once again, it was all about attitude. The top bunk was just fine, and I would make the most of it. The other women asked why I always looked calm and never showed stress from the heat. My answer was brief: There is no use sweating over minor stuff, and I think and pretend like it is 50°F. I thought that 50°F would be a comfortable temperature, considering that it was almost 150°F in Kuwait. I applied sunscreen and wore long sleeves, sunglasses, and a balaclava the entire time I was in Kuwait to prevent sunburn and to be prepared in case of a dust storm.

The ladies' room was across the path from our tent, so I had to get out of the tent every time I needed to empty my bladder. It was challenging to get down from the top bunk and also to climb back up. I thanked God for the steps and for never falling during the nights. It could have been disastrous.

I looked for the dining facility the next day and noticed that it had also changed. A number of tents had been added since the previous time I was here, and I had to go around these tents to find the dining facility. I was surprised that I had to pay for my meals, which was $5.00 for all you can eat. This was quite the opposite of when I was here the last time and had free meals, courtesy of the coalition forces. I came prepared and I paid for my meals, three times a day, and made the best of it.

Hardest Tour

This tour was the hardest one. This might be surprising to hear, but it was the most difficult tour I ever had during OIF because I lost a friend. Later, I found out that my friend had been killed in action (KIA). He was killed in an ambush on his fourth tour in Iraq. I was shocked when I read a newspaper story stating that it would be the last time the American flag would be flown at half-staff for American heroes who were KIA in Iraq. I was intrigued by the subject, so I continued reading the article. I gasped when I read three names and one of them was my friend. The news stated that my friend had been killed in an ambush and was pushed into one of the portable toilets. The enemies took his weapon, his gear, and everything they could find, but they did not know that he had a cell phone in his pocket. According to the news, he called his wife and broke the terrible news to her. The American troops rescued his body, but it was too late. He died, as did the other troops who were victims in that ambush. This was shattering news, and it took me a while to believe it. I remembered seeing three coffins being unloaded from an airplane when I took R&R in Clarksville, Tennessee, in June. As the passengers got off the plane, I saw lots of people looking through the windows of Nashville International Airport. I was curious about the focus of the attention and what they were looking at and saw three coffins on the ground. I silently prayed for the souls of the dead and their families, and then I proceeded to the baggage claim. I had no idea that one of those coffins had belonged to my friend until I returned to Iraq and read the news.

This loss was tragic, and I hoped that I could see someone I knew and talk to them about the tragedy. Indeed, one day in the Camp Striker DFAC, I saw someone I knew who had deployed with the 101st Airborne Division (AASLT). It was Chief Warrant Officer 4 Negron, an officer in charge of the Promotions Section that I used to serve with. I mentioned my friend's death, and he said that he had heard about all that happened to the three victims and that the Army was investigating who did it and all that had taken place. I nodded and said that I would keep praying for him and for the safety of all the troops who were in harm's way. He tapped my shoulders and said for me to take care and that he was glad to see me again. I said the same thing. It felt weird talking to him now that I was a civilian. I truly missed the service in uniform, but I had to move forward. I was told that the deaths of my friend and the other two soldiers were the talk of the town for a while and that a solemn memorial service had been held for the three of them. When he was alive, my friend had served as the non-commissioned officer in charge of Bastogne's personnel service brigade. He attended every S-1 (human resources) meeting, and we always had good conversations together with Major Hargrow. He always came by our office to take care of his troops in their personnel record reviews, evaluations, finance issues, and promotion packets. On the side, he was very active in his community, serving as a pastor and as a sports referee for basketball games. He finished his MBA and continued to serve in the Army. He lived a full life. As I continued to read the news article, I learned that he and the two other victims were the last troops KIA to have the American flag flown at half-staff. My friend's demise was commemorated by naming a street after him.

Later, during one of my visits to Fort Campbell, I saw the street sign bearing my friend's name by Bastogne's HQ. I was driving around to see the improvements made on the base since I had left in 2007. The 101st Airborne Division's HQ had been moved, and the personnel service building occupied the old Commissary building. There was a new PX, as well as new commissary buildings. The soldier processing center was still the same; my old office building was still standing strong; and my son's former childcare building was still there, filled with life. Before I knew it, I was driving through unfamiliar ground. The changes in the area had probably contributed to my getting lost, and I needed to return to my lodging. I looked at the street sign and saw my friend's name. I stared at it for a while and then realized that maybe he had wanted me to be lost and had taken me to the Bastogne HQ to see his new lane. I smiled, and with my silent freedom, I said, "I see your new street, my friend, and may you rest in

PART III: Back to Iraq. Destination: The Zoo

peace." I prayed for his soul in silence, and as if in response, I saw his smiling face, as if he were telling me, "It is good to see you too, my friend." I smiled and started driving back to my quarters. I felt comforted by the knowledge that his memory would last for a long, long time and, indeed, for as long as Bastogne existed.

TRANSITORY FORMATION

My new home at the zoo, Baghdad, Iraq.

My new job as a liaison officer in charge of linguists under a large U.S. company in OIF was fulfilling. I managed the linguists' service records, pay and personnel benefits, assignments, and other information pertinent to their position as linguists or interpreters within the country of Iraq. We had two categories of linguists: U.S. and local. I managed the U.S. interpreters, or "terps," as they were popularly known by troops in the field. My day started at 0800 hours and was supposed to end at 1700 hours, but it was different in the combat zone. Often, the day ended after 16 hours of work. I had a lot of energy during this tour and worked as required. On my first day, I was in formation at 0800 hours inside a beautiful building at the zoo. This must have been Saddam's guest house. It was spacious, and the painting was very detailed and just looked magnificent.

I stood in the formation as a civilian, and it was such a weird feeling. Nevertheless, I slowly learned to get used to it as time went by. The person in charge of the formation introduced herself as Kathy. She called the roll and briefed us on the who, where, why, and what to expect in this assignment. While listening to her, I could not help but admire the details of the building. The facility was far from the destitute landscape of Mosul that I had encountered during the invasion of Iraq in 2003. Although the building appeared to be damaged outside, this concrete facility evidently held its strength against the damages of war. Inside the building were molded high ceilings, marble halls, and beautiful chandeliers.

The formation was an opportunity to meet with the new linguists before they were assigned to the military units in the field. I had met these linguists when we were in training in Herndon, Virginia, and at Fort Benning, Georgia, for more training, safety briefings, medical screening, inoculations, and more activities to prepare for the deployment. Based on our conversations, they had accepted the job for different reasons; some of them said they wanted to support OIF, whereas others said it is for personal reasons. They found out that I was a Veteran and thought that I was crazy to return to the battlefield. I just smiled, but silently I wanted to tell them that I felt at home in this place. It was likely that no one would understand. But they were right: Who in their right mind would give up the luxury of life and peaceful living in the

PART III: Back to Iraq. Destination: The Zoo

United States where nobody was shooting at them and go to a war-torn country like Iraq, where gun fights occurred every day and one could not even go outside the wire? I did not let it bother me, and I continued my rendezvous with destiny.

After the formation, I was introduced to Luis, who would be my supervisor. He looked harmless to me, so I shook his hand, and we exchanged hellos. Luis explained that he came from California. I told him that, because of his appearance, I could have mistaken him for Filipino. He laughed and said he wished he could speak Tagalog. We found common ground when he said that he was a retired command sergeant major in the Army. We could both speak the Army language. Luis spoke calmly and intelligently. After a brief talk, he took me to his office and showed me where I would be working, namely, a building called Z-1 located next to his building. There were three staff members in the office. He introduced me to Bey; Amanda; and Rosemary, who would be my immediate boss. Rosemary's briefing regarding my job was thorough, but a little complicated. Dealing with human capital is not easy, as it entails both professional and personal undertakings, especially in cases of emergency family matters that required someone's presence. There were also reports to be completed and sent to the HQ in the United States. This was all part of the contract. When I asked Luis about the guards at the front gate, he said that the guards were not from Uganda. They were from Fiji. I felt victorious and Luis asked why. I mentioned how other night I thought they were Fijians, but Mr. Arshad said they were from Uganda. Luis gave a good laugh at it and said again, "They are Fijians!" And we both laughed.

WORK CHALLENGE

I liked the work challenge and started working hard, not just to impress my bosses but also because I liked my job. Rosemary was sweet and kind. She briefed me about reporting to work at certain hours and about using the chow halls, the shower rooms, and the laundry room, which were located outside our resident quarters. I was grateful for the warm weather when she said that. I could imagine how challenging it would be if it were freezing cold at night and I wanted to use these facilities. She also said that I should be used to the situation by now since I was a Veteran and had served twice in Iraq when I was in the Army. I smiled as she thanked me for my service and welcomed me back. She asked if I could drive a manual car. Her question took me by surprise, and she read through my facial expression that the answer was no. But I told her that I could drive Humvees and anything else other than a manual car. She said that the only car available at the zoo was a manual and that, if I wanted to go to the other camps, I would need to start learning—and to do it fast. She said that part of my survival would be the ability to drive a stick shift as if my life depended on it, and she said it so seriously that it almost made me freeze, just like my encounter with those savage dogs at Camp Udairi, Kuwait. She added that I needed to eat and use other facilities outside the zoo compound. I asked where I could attend religious services, and Rosemary said there were none at the zoo. If I wanted to go to church, it would have to be at Camp Striker and within the Victory Base Complex, about 20 miles from the zoo. In silence, I thought that I might just be returning to the United States sooner than I had planned. There is no automatic car, and I couldn't imagine walking 20 miles every single day. That would wear me out, and I would always be late for work. Rosemary must have read my thoughts as I chewed on her words, and she said, "No need to worry. Bey will give you driving lessons every day until you can drive on your own." Bey was working in the corner, and at that point, he looked up and almost said, "No way, Jose."

Driving Lessons

For the first two weeks, I depended on Bey to drive us to the DFAC, PX, and other facilities outside the zoo. He told me to watch what he did to start and stop the vehicle and how to shift gears. I did just that as Bey patiently taught me to drive the stick shift in the morning and in the afternoon. The training in the morning lasted about 15 minutes, before we started working at the office. The afternoon training lasted longer and occurred before we went for the chow hall. The first day I showed up for the training, he told me to follow his instructions while watching what he did. He noticed my excitement and laughed about it. On the first day of training, we both walked to the car, which was parked behind Z-1. Once we were seated, he identified the clutch, the brake, and the accelerator. I noticed that Bey was as articulate as Luis. He clearly explained the process to me: The clutch would allow me to disengage the engine from the wheels while changing gears. It should be operated using the left foot. The middle pedal was the brake. The pedal on the far right was the accelerator. Both the brake and gas pedals should be operated using the right foot.

He emphasized that, before starting the car, I needed to make sure that the car was in neutral. The car was in neutral position if the stick felt loose to the touch and could be moved easily from side to side. He looked at me with a stern face as he said this, and I told him to stop and continue with the training. I asked him what would happen if the stick were not left in neutral position. He looked at me and said that as long as the engine was off, then the car would not hit the building. I gasped as he continued. If the shifter was not in neutral, this could be fixed by pressing down fully on the clutch and moving the shifter into the central or neutral position. In answer to my question, Bey said that I could also put the car in neutral by pressing down fully on the clutch with my left foot. I nodded to indicate that I understood, as I did not want to interrupt his train of thought.

He continued with his instructions, explaining that once the car was in neutral, I was ready to turn the key in the ignition and start the car. He cautioned me to remember that if I put the car into neutral by moving the gear shift into the neutral position, I could turn the key in the ignition without having to depress the clutch pedal. I thought, "That's a relief!" He also told me that I did not have to depress the brake pedal when I started the car. However, I thought about that further and asked why not. He said that pressing the brake when starting the engine applies to automatic cars. A stick shift is different from an automatic car. Again, I nodded, telling him, "Roger that, Maestro." We both burst into laughter. It was a good icebreaker after all the serious training about the stick shift. Then, he resumed the instructions, explaining that if I put the car into neutral by simply pressing down on the clutch while the shifter was still in gear, then I would need to hold the clutch down as I turned the key. Otherwise, the car could lunge forward. He further emphasized that to avoid hurting myself or anyone or anything else, I had to seriously memorize those steps. Again, I said, "Roger that," and he instructed me to watch him as he performed lesson one. I saw him touch the stick to ensure that it was in neutral position and then insert the key to start the engine. Once the car was running, he fully depressed the clutch and put the gear shift into first gear by moving it to the left and then forward. The number 1 was clearly marked on the top left corner of the stick. Then, he slowly lifted his foot off the clutch until the engine speed or RPM started to drop and the car began to move forward slowly. He said that this was also called the "biting point."

The Biting Point

Bey explained why it was called the biting point. He said it was the starting gear before the car could be shifted to higher gear. After finding the biting point, it is time to start pressing down on the accelerator, but slowly and gently so as not to kill the engine. I saw that he moved his right foot and started to press down on the accelerator, which was on the far right, next to the brake pedal, while his left foot continued to release the clutch in a simultaneous motion. He told me that, if I completed the action correctly, the car would start to move forward and would be driving in first gear. He was driving slowly through the dirt until the car hit the paved street. The car stopped. Bey was cool. He was such a good driver, and I knew that he was testing me. He asked me why the car stopped. Being a novice, I did not know the answer, and he said with emphasis to beware of stalling. I got it that the driving lessons were serious because my life and the lives of my passengers would depend on me not only in going to the chow hall, but for incidents that required immediate action if we were under attack and needed to drive away immediately to safety or if an employee needed to be transported for emergency medical care.

He continued with the instructions and admitted that he released the clutch too quickly and caused the car to stop. On the other hand, pressing down too hard on the accelerator before the clutch is fully released can wear out the clutch and damage the vehicle. I told him that this would be the last thing I would do because it would mean a long walk to the chow hall and no church service for me. He laughed and said that I was so much fun to teach. He told me not to worry, as I was bound to stall at least twice or three times, and then he said my favorite phrase: "Been there, done that, got a brown shirt to prove it." He said that finding the perfect balance between releasing the clutch and depressing the accelerator would take a lot of patience and practice. I told him that my motivation was to eat and to be able to attend church services. He laughed again and then showed me how to shift into second gear. He told me to shift my eyes onto the stick and listen closely. When the engine started to race and sound like it was under pressure, approximately 2500 to 3000 RPM, depending on the car, then it was time to move up into second gear. This was to be done by taking the right foot off the accelerator if necessary and using the left foot to fully depress the clutch. Then, grab the shifter and move it straight backward into second gear, which should be marked with a number 2 on the stick.

Bey mentioned that there were rough hills out there in Camp Slayer, so he showed and explained hill starts to me. To prevent the car from rolling backward, he said to begin with the left foot depressing the clutch and right foot depressing the brake. I looked at him, asking if you can really depress both pedals. He confirmed, saying "Affirmative." To stop rolling backward on a hill, he instructed, depress both the clutch and brake pedals, put the car into gear, release the hand brake, and then lift the foot off the clutch until the biting point is found. Then, release the brake and press down on the accelerator, using slightly more gas than usual, and then continue driving as normal. I asked why he would want to keep the clutch at the biting point to perform a hill start. He stated that keeping the clutch at the biting point prevents the car from rolling backward. I agreed that that made sense—unless it was the enemy car, and then I would release both the brake and clutch pedals to knock them down. "Smart answer," he said, and laughed again. One more lesson, and then we would be done for the day. He added that to perform a hill start using the hand brake, you should put the foot on the clutch and put the car into gear. Slowly release the clutch

PART III: Back to Iraq. Destination: The Zoo

until the biting point is found, and then release the hand brake. Once the hand brake has been released, put the foot on the accelerator and proceed as normal.

Bey asked if I had any questions about the driving lessons today and told me that I did not have to raise my hand since I was his only student. It was my turn to laugh. I asked him to explain once again about the biting point. Good question: The biting point refers to the point in the clutch pedal's travel where you feel the car start to move. My second question was, is there a quick way to restart the engine, especially in case of emergency? We were in a hazardous location and reacting quickly and in the right way would save lives. He answered that I should always remember not to release the clutch too quickly, because that would make the car stall, and I would need to start the process again. I also asked how to stop the car when it was necessary to stop right away. He responded with a grin and said to push the clutch in all the way and then press the brake. Then, when I wanted to get moving again, I should take my foot off the brake, slowly depress the clutch until I reached to the biting point, and then press the accelerator pedal to start moving again.

I was about to ask another question, but we had reached the work site, and he parked the car and told me to ask the other questions inside the office. However, I never brought up my other questions on that day because I had to focus my attention on my work. Rosemary asked about the progress of the driving lessons, and I answered, "Great! Bey motivated me that I cannot eat unless I learn how to drive!" Rosemary burst out laughing and said that that was largely true and that I must learn quickly; otherwise, I would starve. We all ended up laughing. We all got along pretty well because of our good senses of humor. I believe that having a good sense of humor is a sign of a healthy mind and body; it is a panacea for boredom, frustration, and fear of the unknown.

I remember the words of my mother-in-law when she was healthy: "A good sense of humor is healthy for heart and mind." And she lived to be almost 90 years old with that good attitude before passing away.

Masterwork

One day, Bey tossed me the key and said that it was my turn to drive for lunch at Camp Striker. My hands went cold, and I looked at him intensely: "Are you serious?" He said that if I wanted to go to lunch, then I needed to drive. I felt nervous, but I am a Soldier for Life, and I had to demonstrate my confidence. It was time to put the driving lessons to the test. I took the driver's seat and stared at the stick. I tried to move it to each side. The stick was loose, which meant that it was in neutral position. I inserted the key into the ignition to start the engine. Great. It was working so far, I told myself in silence. Bey was pretending to be looking on the other side and ignoring what I was doing with the car. It would have been easier if he had just gone ahead and driven while I took the passenger seat and looked somewhere else, just like he was doing right then. But instead, it looked like he was going to trust me and believe that maybe I was capable enough to drive us both to Camp Striker. That was a strong assumption.

The car engine finally started, and I wiped a bead of sweat from my forehead. The next step was to depress the clutch and move the stick into first gear. I was about to drive forward when the car stopped, and we lunged forward. I bit my lower lip and did not want to look at Bey, although I could feel his stare as if saying that I let the clutch go too quickly. He told me to try it again. At that time, Rosemary came out to join us on the trip to Camp Striker. I felt more beads of sweat, this time in my armpits. I think Rosemary saw my jaw drop, but she tried to ignore it. With my silent

PART III: Back to Iraq. Destination: The Zoo

freedom, I came to believe that they had both ganged up on me that day to watch how I would react. I was hoping that Luis would not join us too, so that I would not be even more embarrassed. So, I tried again, and this time, I slowly lifted my foot from the clutch until the RPM began to drop and the car started to move forward. I started heading toward the concrete path when the engine stalled again. This time, I looked at Bey, who told me to feel the biting point and then press on the accelerator slowly so as not to kill the engine while simultaneously releasing the clutch. The car would then drive in first gear, and then, I would change the gear as it moved faster. I said, "Yes, sir," and I heard Rosemary chuckle. I gave a big sigh, and with confidence, I told myself silently that I could do it, that my life depended on it, and that my passengers were counting on me. I repeated the process, and this time, I was hoping to feel the biting point, as I slowly lifted my foot off the accelerator and depressed the clutch just enough to shift from first to second gear.

It felt surreal, but I did it. My feet were moving like a pro, and I was changing the stick shift into different gears as I was supposed to be doing. Rosemary said that I had perfected the mission and mentioned a litany of freedoms I could enjoy with my new masterwork. She said that I could now live a sophisticated life, sophisticated enough for the zoo. I smiled as I saw her excitement too. She continued and said that I could go outside the zoo, eat at fancy DFACs, attend religious services, drink good chocolate mocha at the Green Beans Café, visit the Al-Faw Palace, go to the International Zone, see Saddam's damaged Victory Over America Palace at Camp Slayer, visit the hadji stores at Camp Victory, and see other limited luxuries. Listening to Rosemary's litany of incentives was good enough to boost my newly found driving skill. Because everything was free at the DFAC, she said she could not buy me a drink. However, she was quick to correct herself and happily said she could buy me a cinnamon bun and good mocha from the Green Beans Café. I was ecstatic, and Bey looked at Rosemary and asked what she would get for the good instructor. We all laughed, and Rosemary finally decided to treat us both. "That's what I'm talking about," said Bey, as he gave a big, wide grin showing his white teeth, but not bad like Drill SGT McCroskey's Joker grin.

Camp Striker

Rosemary told us a brief story about how Camp Striker originated. Camp Striker was one of several logistical and life support bases with the Victory Base Complex, Baghdad, Iraq, near Camp Victory. It was established in 2003 by the 2nd Brigade, 1st Armored Division, known as the "Strike Hard" Brigade. Their home base is Fort Bliss, Texas. Rosemary further said that we actually owed them for the freedom to eat at the second largest DFAC in Baghdad, as well as the other amenities they instituted before they left Iraq. It was fitting that the camp was named after the Strikers because of their resolve to improve the quality of life for Iraqi citizens and implement stability operations in Baghdad and its neighboring towns. They built capacity and security for the Iraqi people to allow for economic and governance development. They built the walls and roads that secured freedom inside the wire. As we enjoyed the fruits of their labor, we were grateful for the triumphs of the brave Strikers in decreasing the prospects of danger and securing the bounties of liberty. Freedom is not free, and they also lost some of their soldiers in this menacing country. General Petraeus, Commanding General (CG) of the Multi-National Forces, Iraq, applauded their efforts and stressed that the work they did was extremely vital. The CG gave out 50 coins, tokens of appreciation for a phenomenal job, to deserving soldiers. Before Rosemary had finished her story, we saw the road sign leading to Camp Striker.

The Road to Camp Striker

A sign pointing to Camp Striker, Baghdad, Iraq.

I followed the sign and passed through the guard checkpoint. We saw another sign showing the Camp Striker directory. The sign was built by the 54th Engineer Battalion, TF Dagger, from Bamberg, Germany. I thought that the signs were well built and informative as they helped visitors locate the different facilities within the camp. We found the DFAC easily and parked the vehicle on a hard spot. We always had to remember where we parked so we could easily backtrack in case of emergency, which should always be expected in dangerous places like Camp Striker. After showing our badges upon entering the DFAC, which had a good crowd as usual, we waited in line for the next available sinks. Washing hands before entering the mess hall is a big tradition in Iraq, practiced on all military bases. By the sink were soap dispensers and several brown paper towel dispensers for drying our hands. We then entered the mess hall and proceeded to the chow line. We let Rosemary lead us as we grabbed food trays. I saw hamburgers, hot dogs, corn and beans, spaghetti, and many more options. I selected spaghetti and meatballs, with corn muffins and an ear of corn. It would have been perfect if they had had corned beef and cabbage, my favorites. Bey chose to go with three hot dogs in buns, corn, beans, and a bag of corn chips. Bey was about 6'2", and I expected him to eat much more than a petite person like me could eat. We snuck a couple of corn chip bags and sodas for snacks later. My cargo pants worked well in the battlefield.

Directory of Camp Striker, Baghdad, Iraq.

The U.S. Army spared no expense in supporting its soldiers. Rosemary mentioned that, even in Vietnam, ice cream was flown to the forward operating bases. The new expanded

DFAC at Camp Striker was served by KBR and soldiers of the 2nd BCT, 10th Mountain Division, out of New York. Camp Striker's old DFAC was a wooden building with several trailers attached to it that had been intended to serve the soldiers for six months; instead, soldiers had used it for more than three years. The previous DFAC could not handle the growing number of troops, hence, the expansion. The new DFAC was expansive, which was crucial for serving more troops and civilian contractors in view of the ongoing surge in the Baghdad area. And here we were, as part of that surge.

The DFAC warrant officer revealed that the facility intended to expand its menu even more, adding to the potato, pizza, and pasta bars with stir-fry gyros, fruits, and international food. To provide more food faster to the troops in the battlefield, the DFAC employed about 300 workers including soldiers on the DFAC staff, American civilians, and third-country nationals (TCNs) from places such as Sri Lanka. The main chef, from Antwerp, Belgium, could speak four languages, but none of the workers spoke these languages. The DFAC kitchen seemed to be a well-placed reference to Iraq's historical Tower of Babel because the employees spoke different languages: Hindi, Urdu, Tagalog (my native language), Nepalese, Tamil, Bahasa, and Arabic vied with the U.S. military standards of English and Spanish. However, the employees still knew enough English to serve the troops and accomplish their missions.

The DFAC was indeed like the Tower of Babel. According to the Bible (Genesis 11:1–9), the Tower of Babel is an origin myth meant to explain why the world's peoples speak different languages. The structure was built in the land of Shinar, or Babylonia. According to Genesis, the Babylonians wanted to make a name for themselves by building a mighty city and a tower with its top in the heavens. The story further states that God disrupted the work by so confusing the language of the workers that they could no longer understand one another. The city was never completed as a result of the disruption, and the people were dispersed over the face of the earth. One of the historical structures associated with the Tower of Babel is the Ziggurat of Ur, which I had the opportunity to tour while I was at Camp Adder.

Ziggurat of Ur.

The Ziggurat of Ur is a 4,000-year-old structure. During OIF, soldiers were given the opportunity to tour the Ziggurat to learn about its history and build a camaraderie among the group. Most soldiers would otherwise never get a chance to see something like this. According to archaeologists, the Ziggurat was completed in the 21st century B.C. and served as an administrative center for the city of Ur. Some believe that Ur was the largest city in the world at one point.

All of Noah's descendants spoke a single language. As they multiplied and increased in number, they began to spread eastward until they found a fertile area call Shinar and settled there. They then decided to build a city with a tower that "reached to heaven." They wanted the tower to be a proud monument to themselves and a symbol that would keep them united as a powerful people. "But the Lord came down to see the city and the tower the people were building. The Lord said, 'If as one people speaking the same language they have begun to do this, then nothing they plan to do will be impossible for them. Come, let us go down and confuse their language so they will not understand each other.'" (Genesis 11:6–7).

That is the Biblical explanation for how people have different dialects and languages, so that we will encounter challenges in things that we do. It was fascinating to me that the historical Tower of Babel was very much present in the DFAC. My silent freedom wanted to seek further information about the Tower of Babel. I was enthralled with its history. Babel was the Hebrew name for Babylon, which means "gate of God," and is similar to the Hebrew word "balal," which means to confound or confuse. The Hebrews abhorred the Babylonians, and there was a humorous play on words that Babylon was far from the "gate of God." Instead, Babylon was actually the site of much confusion. At least at the time of our visit, there was mutual intelligibility, and the DFAC kitchen employees could mutually speak and understand each other by speaking English. Another approach to effective communication is sign language. Oftentimes, Bey and I used sign language with the TCNs to get the right blend for good smoothies, for example. We pointed at the fruits we wanted and gave a thumbs up if the smoothie was good. We never gave a thumbs down; they were good at what they were doing, and we kept it that way to make friends.

After learning the history of the new DFAC, my eyes roamed around to hopefully see familiar faces. It was a strange feeling to be a part of OIF again, especially as a civilian. I had not quite gotten used to it yet. I noticed that it did not take a long time to fill the tables, and before we knew it, some soldiers and civilian personnel joined our table, enjoying the food, peace, and freedom the battlefield could offer. A few minutes later, I noticed that some civilians, specifically, several gentlemen wearing neckties and two ladies, had entered the mess hall. I concluded that they must have been part of the Government Accountability Office and that they had come all the way from Washington, D.C., to perform surveys and maybe report to Congress on the progress the U.S. troops and contractors were making in helping to improve the Iraqis' quality of life. My eyes were busy scanning for a familiar face from 101st Airborne.

I had read that the 101st Airborne Division would be deploying to Baghdad at this time of the year, and I could not wait to see some of my comrades who were still actively serving at Fort Campbell. I made a note to remind myself to research whether they were in theater now or maybe later. I recalled that it was in late January when we deployed for the first time in Mosul and at the end of summer when we deployed the second time to COB Speicher in Tikrit. The memories were so vivid. The memory of the Green Zone in Baghdad during the first deployment was brief, as I passed the busy traffic of soldiers with different unit patches moving back and forth with their weapons. It was the day my St. Michael had picked me up from Camp Udairi to Baghdad, a stopover to Mosul to rejoin my unit. I also recalled seeing many helicopters with their rotors on while they were dropping off soldiers. Those memories were from the invasion and chaos in 2003. What a big change since then, and I was so glad to witness all of it. My silent freedom was screaming that we were back in the battle zone and that we were doing the same thing again.

PART III: Back to Iraq. Destination: The Zoo

Bey interrupted my train of thought by asking if we should get more dessert for the people at the office. The three of us stood up and went for the cookies and apples, in addition to the bags of chips already in the pockets of our cargo pants. Bey mentioned that we needed to stop at the fuel stop point and get gas. Before I could open my mouth and ask where it was, he said to turn on the engine and he would guide me. Because I was still quite a novice, we stalled again twice before the car was in motion. It was embarrassing but also part of the learning curve, and I was on cloud nine as I finally learned to drive a stick shift, which I found to be addictive later. I had been salivating to drive a stick shift at home, but I never got to it. What a shift of fortune. I had to learn how to do it in the battle zone, but it was the only mode of transportation and how we could partly survive. Bey guided me as I reversed the vehicle and made sure not to hit anyone or any other vehicle. I was so happy with my newfound knowledge of driving a stick shift, and Bey noticed and kept teasing me about it.

There was a long line of vehicles waiting to get fuel when we reached the fuel point. Rosemary said we would just have to be patient, get in line, and top off the tank. Bey said, "Aye, aye, Captain," and we all laughed. Rosemary was such a sport and so easy-going. We could be ourselves and did not have to change our behavior or pretend like we were different people in front of her. She got it and said that we were one team, one fight. When we finally got our turn at the pump, a staff member handed a clipboard to me, and I filled out the form and signed my name on it. When we had first gotten in line, another staff member had told me to stop the engine, which I did. Getting fuel at Camp Striker was not difficult; it was going through the check points before entering the camp that was challenging, as the guards asked us to show our badges and answer probing questions. Given that we were at war, this was completely understandable. We could be infiltrated by the enemy trying to blend in with us. Checking badges and asking questions might help identify the bad guys and prevent them from invading the camp with their homemade IEDs. The thought of having a spy inside the wire was a frightening thought. A suicide bomber could be sitting next to any one of us and blow us all up. It was not easy working in a country in conflict. We always lived on the edge and as if walking on a thin wire, not knowing if it would be our last time to breathe. There was no time to say goodbye to our loved ones. Anything could happen in the battle zone.

False alarms were nothing unusual, as we had gotten used to them. We ducked and took cover and waited for an all-clear signal. Dust storms sometimes resulted in false sensor reading. We did not try to hide anywhere else during an alarm; we took cover. Fortunately, contractors went through security and safety briefings and also knew how to react in times of close calls. My best weapon was my faith in God, which helped me avoid feeling complacent and take cover in fortified locations. It would have helped if I had been armed with an M4, but as contractors, we were not authorized for such weapons. Once again, I appreciated the concrete blast walls surrounding the compounds. They proved to be effective barriers by providing security and protecting us, establishing stability, and eliminating terrorist threats. They were our protection from indirect fire from rockets and mortars. They had bunkers and guard towers. Rosemary mentioned news about the growing numbers of IEDs placed on top of the concrete T-walls. The terrorists used advanced forms of IEDs, suspected to have come from foreign sources such as Iran, that could penetrate any blast walls. This allowed modern forms of IEDs to be placed on the non-road sides of the barriers. However, the concrete walls did take away the ease of access for terrorists to emplace

IEDs, degrade the lethality of their homemade devices, and force them toward specialized materials that could be prohibited at checkpoints. The concrete walls were highly effective and were used to convey traffic through channels and thwart the emplacement of IEDs. The concrete walls took away the ability of insurgents to freely transit Baghdad with large, vehicle-borne IEDs, which created mass casualties and threatened the authority of the coalition forces and the Iraqi government. Such concrete walls were our life savers. Beyond the blast walls, I maintained my faith in the Lord, my Savior, and entrusted Him with my life, especially in the battlefield. For me, it was a miracle to have survived the hostilities in Iraq.

Linguist or Translator

We returned to the zoo and continued working on issues related to our linguists assigned to support OIF in outlying units. I strongly believed in this mission, as our translators bridged the gap between the Iraqis and the coalition forces. Our linguists were the stronghold, the rock, the bastion, who courageously helped by interpreting and speaking Arabic, the country's language, to support battlefield commanders while stationed in Iraq. The coalition forces increasingly came to rely on the contractors to provide linguist services to support effective functioning and facilitate improved communication. The services of our linguists were crucial in every corner, anywhere our troops were actively engaged. Technically, there were two primary types of linguist services: interpretation and translation. Our linguists offered both services. However, although our contractors did both, interpreting and translating are two different professions. Interpreters deal with spoken words, whereas translators deal with written words. Also, whereas interpreters often work in both directions from and into two different languages, translators generally work only into their active language.

The terps I managed converted English into Arabic or vice versa, or in the case of sign-language interpreters, they converted between spoken communication and sign language. This required the interpreters to pay attention, understand what was being communicated in both languages, and express thoughts and ideas clearly. It was also important that the interpreters have strong research and analytical skills, mental dexterity, and an exceptional memory. Our embedded interpreters travelled with our armed forces. It was important that the interpreters see the communicators to hear and observe the person speaking and to relay the messages from and to our soldiers.

The translators in Iraq converted written materials from Arabic to English, so they had to have excellent writing and analytical abilities. The documents that they translated had to be as flawless as possible, so the translators also needed good editing skills. When the translators first received text to convert into English, they usually read it in its entirety to get an idea of the subject. Next, they identified and looked up unfamiliar words. Multiple additional readings were usually needed before the translators began to write and finalize the translation. If the translators were unclear about anything in the text, such as unfamiliar ideas, words, or acronyms, they also would do additional research on the subject matter. We had two different types of terps: the U.S. interpreters, which we hired from the United States, and local nationals. Bey and I took managed the U.S. interpreters, whereas the local nationals were taken care of by another team in our company. All terps had to go through academic and medical screening; oral and written testing processes; and on top of it all, a background check.

Camp Victory

When I had been at the zoo for a month, Bey announced that because I was driving like a pro, we were going to have lunch at the Sports Oasis DFAC at Camp Victory. I did not have to show my excitement. Bey already knew how ecstatic I was to hear that we were going to Camp Victory. Rosemary could see the new twinkle in my eyes, and she also joined us. I kept repeating to myself that it was a dream come true to be able to drive a stick shift. Camp Victory was about 10 miles farther from the zoo than Camp Striker. The Camp Victory was huge and had a lot more activities to offer. It had a small Army & Air Force Exchange Service (AAFES) shop south of the Sports Oasis DFAC, along with a Green Beans Café, a Pizza Hut, a barber shop, and a Turkish novelty goods stores. As we roamed the area, we saw that the camp also had two basketball courts, one of which had been converted into a soccer court. Between the residential chus and the eating establishments was Tumlin Field, a popular spot for American football pickup games. The Tumlin Field sign read, "Tumlin Field, 'cause not all the fighting is done outside the wire." I truly admired their wit. It was a good outlet for the troops, as all work and no play make GI Joe and GI Jane dull soldiers. Rosemary liked the sign too. She loved and fully supported the troops.

We found the trailers of Dodge City North. I told Bey and Rosemary that not too long ago, I had lived in one of those chus. They were small but convenient and much better than sleeping in tents. The chus could withstand much abuse. They were shipping containers that had been modified with hard floors, windows, and air conditioning. Each chu could house up to three soldiers. The one I lived in could house only two soldiers. Most of the chus were equipped with beds, refrigerators, and antennas. The soldiers had to pay for the antennas. The antennas converted radio frequency into alternating current or vice versa, and both receiving and transmission antennas resulted in better TV entertainment because they provided better reception. We called our living area chuville, because there were so many chus, fortified by sandbags and a tall, concrete wall.

Some of the modern chus have built-in showers and toilets. During my time, the chus did not have these luxuries. Our showers and toilets were located outside, and I thanked God for the warm weather. I did not have to wear a coat or hat. I heard that they also came with refrigerators and TVs. Sometimes, chus were stacked two high and lined up in neatly organized rows. Chus were often surrounded by sandbags for protection. One thing we had in common was the tall concrete walls surrounding the chus for security and fortified protection against blasts and enemy attacks. The fortified chuville allowed more freedom for the troops, as they knew that they were protected by tall, impenetrable walls. Helmets, Kevlar vests, and guns were ready if we were attacked. Fortunately, the enemy had not even dared to get close to the fortified chuville during my time there. I cannot imagine the outcome if they had.

The trailers of Dodge City North. The tall structure in the distance is a Saddam-era bat house.

PART III: Back to Iraq. Destination: The Zoo

I was not sure if I had impressed Bey with my story or if he had never had the chance to live in one of the chus because he did not say anything. But he agreed that they looked like they would be cool in the summer and warm in the winter, except that there is no winter in Iraq. The weather in Iraq is like the fall when it is cold. As we moved on, we saw that Sports Oasis was also surrounded by T-walls, and although wearing soft caps, soldiers and civilian personnel were always wary of anything unexpected; after all, we were in a battle zone, and we could not afford to be complacent. We were always on alert and wary of our surroundings. We were required to carry our helmets and Kevlar vests in case of RPGs or an insurgent attack. It was mandatory and a lifesaver. No matter how inconvenient it was to carry the heavy armor, it was better for us to be safe than sorry. Additionally, it was the company policy to comply with existing battlefield rules.

We were even trained to park our vehicles tactically, called combat parking, for swift egress and evasion. Bey explained that swift egress and evasion meant that by pulling forward and reversing into a parking lot, we could quickly leave our location. It would be significantly faster to jump into a vehicle, start it up and immediately pull out safely, compared with having to reverse out into the traffic. It was dangerous in every corner of the battle zone. We always faced incoming threats, so we always tactically parked our vehicles, a habit that I find safer even to this day. When backing out of a parking space, my field of view looking through a rear-view mirror is severely limited and can cause me to miss oncoming traffic.

I say this speaking from a bad experience. One day after I had returned from my first deployment, I hit a school bus when I was backing out. I was stunned, stopped the car, and talked to the bus driver, who admitted that he had seen me backing out but had not stopped. I asked why he did not stop when he actually saw me backing out, and he did not answer. The incident caused a dent in my trunk, but no damage to the school bus. It was an experience that I will remember for a long, long time and that still causes me to tactically park my vehicle, any vehicle, anywhere, even shopping and parking at the mall. I find it more convenient, as it provides a seamless exit into traffic. It might take relatively more time to reverse depending on the cars on either side or behind the car, but I find it easier to align my car parallel to the lines in a parking spot, and I find it safer to pull forward and avoid the risk of getting hit or hitting another car. It has been my bad habit to tactically park my vehicle even when I am not in combat, although some people say it is a good habit. Lately, I have found more people parking their cars the way I park.

Thanksgiving at Oasis, 2007

That year was my first Thanksgiving celebration as a civilian contractor in Iraq. As Bey, Rosemary, and I entered Oasis after washing our hands, we saw how beautiful the Thanksgiving decorations were. Some were hung on the wall, and a paper turkey cutout was on top of a giant multilayer cake that read Happy Thanksgiving. They had divided the word Thanksgiving into two words, and it actually looked good on a layer cake. The cake was at the center of the DFAC on a white table, draped in gold satin fabric. It was a mesmerizing view, and the three of us joined the chow line for turkey with all the trimmings.

Again, I scanned the room for familiar faces from the 101st Airborne Division. I noticed familiar unit patches, and I felt like I was serving with the 101st Airborne Division again. Bey interrupted my thoughts and guided both Rosemary and me

Oasis dining facility filled with Thanksgiving decorations.

to the table. The turkey, stuffing, mashed potatoes, sautéed green beans, and everything else looked just great, and the decorations were enough to brighten up our meal. The chefs really armed themselves with good recipes for Thanksgiving. We noticed that the salad bar was always filled with fresh vegetables such as lettuce; tomatoes; carrots; cucumber slices; broccoli florets; red, orange, and green bell peppers; red and yellow onions; cranberry slices; bacon bits; the best-ever cranberry sauce; and many toppings, with a good selection of salad dressings and a variety of cheeses. I told Bey and Rosemary that if the food was always this good, I would stay in Iraq. Rosemary told me never to say that curse again. Bey laughed at my sudden silence.

  I focused my attention on my full plate, and we were amazed by the food management's efforts at Oasis. They even baked a variety of desserts and edible decorations. Desserts included chocolate mousse cake, pumpkin pie, carrot cake, vanilla cake, and many others. Near beers also stood out in the fridge, including O'Doul's, Coors, and some German near beers. We had to comply with General Order No. 1, according to which the U.S. Army bans the use of alcoholic drinks in the combat zone. Even though we were civilian contractors, we had to abide by the rule because we were in the combat zone. A side effect was fewer alcohol-related disciplinary problems. No alcohol meant that there were fewer cases of troops getting in trouble with local civilians. There were also far fewer brawls, murders, and rapes. I recall a report that there was a decline in alcohol-related problems throughout the force after the combat-zone prohibition went into effect. At the same time, there was a spike in alcohol-related problems in units that had returned from Iraq, as some troops tried to catch up on missed drinks. Some troops did not realize that it had become a problem until they met with their physician, who declared that they were addicted to drinking alcohol. Catching up on missed drinks was also explained as an attempt to cope with the stresses related to serving in a combat zone because everyone is stuck in the fighting field and dealing with family problems at home. Nevertheless, the months without access to alcohol helped many troops learn how to do without it or to get by with a lot less.

Witnessing the soldiers and civilians celebrating the Thanksgiving holiday in Baghdad, Iraq, brought tears to my eyes. I recalled sitting at the table with my team in the previous year as we sat at the same table. It was rare that we ate together, but we did on Thanksgiving. It was like yesterday, eating pumpkin pies in Tikrit and exchanging life stories. We were a big family deployed in the middle of the desert. We were all wearing our Army Combat Uniforms (ACUs) and sitting comfortably as we enjoyed each other's company for about 20 minutes. The ACU is the current combat uniform worn by the U.S. Army and U.S. Air Force. This combat uniform was the successor to the battle dress uniform and desert camouflage uniform worn from the early 1980s to the mid-2000s and from the mid-1990s through the early 2010s, respectively. I thanked God that the soldiers in this deployment were still wearing ACUs, the same type of uniform as I had worn, which made me feel that I still belonged, which for me lasts for a lifetime. I was also missing

PART III: Back to Iraq. Destination: The Zoo

my M4, which was my weapon buddy during the second deployment in Iraq. I had not given up on trying to find familiar faces as Bey, Rosemary, and I had lunch. I thought that maybe if I frequented Oasis, I might finally encounter some of the familiar faces that I had been searching for—faces and Screaming Eagles combat patches from the AASLT.

Bey and I discussed Camp Victory and how it got its name. He stated that what Americans knew as Camp Victory was known by the locals as the Al-Faw Palace, one of Saddam Hussein's masterpieces. He explained that Camp Victory was named after V Corps, which was also called Victory Corps, based in Heidelberg, Germany, which began to occupy the area in April 2003. Camp Victory had several living support areas, namely, Freedom Village, Dodge Cities North and South, Omaha Beach, Audie Murphy LSAs, Red Leg LSA, and the Brickyard along with building 51F, which was commonly known as "Area 51." There were also two smaller living areas reserved for government contractors, as well as a third for employees of an Iraqi contracting company. Then he looked at me and said that most of all, the complex had Camp Slayer, which was known for shops selling souvenirs from the country that were good to visit if one was looking for bargains. I asked him how he got all that knowledge. He responded that he had been reading about it, and he gave a short laugh. Rosemary said that it pays to read when one is living the history. I remember how I looked up to Rosemary with so much respect when she said that.

We spread the good news to the rest of our crew as soon as we got back to the zoo, and I gave them the keys so they could take their turns and have their Thanksgiving meals, too. It was a good Thanksgiving meal away from home, in a battle zone supporting the troops. We could only eat so much and could not wait until the next holidays: Christmas and New Year's Eve. I told Bey that I would like to go to the

DFAC at midnight on New Year's Eve, but he said that would be not a good idea in this area because of the distance we would have to drive. It was not like Kuwait, which was safer. I thought that this was a valid justification but suggested that maybe we could go to the Special Forces Group (SFG) DFAC. I remember Bey telling me that he had always used the DFAC at the SFG compound. It was quiet and close to the zoo. It was walkable from the zoo and had a gym and cleaning service facilities. It was a beautiful complex run by contractors from different countries. According to a Congressional Budget Office (CBO) study from 2003 through 2007, the following countries were part of the Iraqi theater of war: Iraq, Bahrain, Jordan, Kuwait, Oman, Qatar, Saudi Arabia, Turkey, and the United Arab Emirates. The CBO found that in-theater contracts to support operations in Iraq were almost entirely performed in those countries, all of which were located within the U.S. Central Command's area of operations. From 2003 through 2007, U.S. agencies awarded $85 billion (in 2008 dollars) in contracts for work to be principally performed in the Iraqi theater, accounting for almost 20% of funding for operations in Iraq. More than 70% of those awards were for contracts performed in Iraq itself.

There were numerous dining facilities in Baghdad, including one in Victory Base Complex, or VBC. VBC was an amalgam of military installations around Baghdad International Airport (BIAP). The complex had 10 bases, including Victory Fuel Point, where we refueled most of the time; Camp Slayer; Camp Striker; Camp Cropper; Camp Liberty; and Radwaniyah Palace, otherwise known as the Al-Faw Palace, which was used as the headquarters and was where the U.S. Armed Forces commanding general resided. There were two man-made hills and several man-made lakes, and the VBC housed two presidential palaces. The northern one, called the Al-Faw Palace, was then occupied by U.S. forces. The southern one was known as the Victory Over Iran Palace, which I would visit during my last tour to Iraq. Bey and I found a third palace with a large and beautiful swimming pool sitting on the hill that housed the SFG. According to hearsay, it was a palace that Saddam had built for his mother.

WALKING BUDDIES

Bey and I became walking buddies by following a route from the zoo to the hilltop in the SFG compound, although he was late most of the time. I had to wake up at 4 AM, and with a bottle of water in my hand, I started my walk toward the palace on the hilltop. I went through a tall gate and showed my ID to the guards, and they let me enter the camp. They became familiar with my face as days went by

The palace on the hilltop that Saddam Hussein built for his mother that was occupied by the SFG during OIF.

and from walking every other day inside the camp. Before stopping at the gym, I walked up the hill and around the palace but never went inside it. How Special Forces operated inside the building was their business, and I had nothing to do with it. It was their building now, and I prayed for their safety. After all, we were one team, one fight.

Palace on the Hilltop

The palace on the man-made hilltop was for the exclusive use of the SFG, and we were forbidden to even peek inside, except for a one-time scheduled medical inoculation with the medics. We were shown the medical office and did not go beyond the clinic to poke around. Too many secrets. I saw some of the Special Forces warriors only when I went to their DFAC and when I used their gym, which was located downhill and outside their palace. After using some light weights and the sit-up bench, I then started back toward the zoo. It was about two miles round trip. One morning, Bey showed up as I was returning to my building and asked why I was always early. I told him that it was one good trait I learned while I was stationed in Korea. We were housed so far away, about a one-hour drive from Camp Henry, Korea, where I worked, and I always woke up at 4 a.m. to catch the first bus at 6 a.m. The next bus was scheduled at 8 a.m. and would make me miss the PT formation and be way too late for work. It became a weekday routine, and I got used to it and did not want to give up such a good habit. I got things done before everyone else showed up at work. I could focus on other worries of the day since I had already finished all of the reports before sunrise. Bey said to wait for him next time so that we could walk together. He said that it was dangerous for me to walk by myself and that I should not walk alone. He was right, but I trusted the SFG. I am a little bit stubborn, and I felt so much better when I started walking early and returned to the zoo before the sun rose.

One morning, I found Bey waiting on the roadside at the zoo. I asked who it was that he was waiting for, and he smirked. He asked why I was carrying a bottle of water. It was my turn to sneer. The bottled water was our topic all the way up to the arch gateway, and we changed topics when we reached the gym. Bey had long strides, and I struggled to keep up with him. It reminded me of when I used to walk with my former husband. He was 6'2", and often, I had to dash to catch up with him. I had to ask him to slow down when I got tired of running just to catch up with him. He stopped, held my hand, and pulled me along so that we could keep the same pace. Those were good memories. We liked doing physical fitness even when we were on vacation from work. My thoughts were interrupted by a question from Bey. We started talking about our families at home and the reasons we were working in Iraq. I was glad when we finally reached the gym and he stopped asking more questions. I pulled out one of the mats and started doing push-ups. Bey went to a bench to lift weights. We stayed at the gym until we noticed more people coming in so the SFG could use the treadmills, weight benches, and other equipment. Bey said that we had had a good workout and that we were just guests, so we gave the SFG their space and we started heading toward the palace on the hilltop. The olive trees on both sides of the street added drama to the path leading to the palace.

Olive Trees

The olive trees were so amusing to look at. Bey tried to pick some olives for me, but he stopped when he saw a man coming our way. We resumed walking and almost ran until we could not see the man anymore. We both laughed at our juvenility, and after catching our breaths, I told Bey that we should be glad the man did not have a gun. It must have been one of the gate guards or one of the caretakers. I told him that we could have been killed for stealing a few olives. He sneered and said that no one was going to shoot at us. After all, our very own Special Forces dominated the area. He seemed so confident when he said that, so I went along with it. There are two olive harvests a year, one in February and one in November. The olives are separated from twigs and leaves and then pitted and squeezed to produce a thick, aromatic oil. Locally, the oil is used in cooking or drizzled on top of favorite appetizers such as hummus. It can be used also to make soap, and the dark, sawdust-like residue from olives pressed in the fall is often burned to heat houses in the winter. I noted that it was almost November, so that must have been why there were so many olives on the trees. We tasted one of the olives, but it was sour, so Bey decided to keep them in his pocket for later. I did the same thing.

After walking four laps around the palace on the hill, we decided to head back to the zoo. I mentioned to Bey that there were date trees next to building Z-25 at the zoo and promised to show him when we got back. We used another path for the return trip to avoid the man in case he was waiting for us. Bey reminded me not to be afraid, and we continued running. I gave a big sigh of relief after we passed the gate and reached zoo land. Bey suddenly burst into laughter, and the more I asked what he was laughing about, the more he guffawed. I told him that it was good to exercise his jaws early in the morning, and he continued laughing. He said that we were like naughty children stealing a fistful of olives and could be thrown into the animal zoo if we were caught. I agreed, except amending that the zoo did not have animals in it, but people. He laughed again.

PART III: Back to Iraq. Destination: The Zoo

SPECIAL FORCES GROUP'S COMPOUND

The palace of Saddam's mother was gated by a wonderful arch. It looked dingy when I saw it, but I was quite sure that it shone like the sun and was well-tended before the war. Later, after Bey departed Iraq, I took a picture of Imee wearing her giant ACU hat as we both walked toward the arch. Neither one of us had the opportunity to look inside the palace, as it was also forbidden. About one mile from the arch was the SFG's DFAC, where Bey and I ate quite often because it was the closest to the zoo. The hall was always filled with Special Forces warriors, and the chef and staff were friendly and accommodating. They always served a variety of good food, such as chicken kabobs, steak, lobster, Mexican foods, and other savory cuisines that made us return for more whenever we were allowed. But we could not go back for seconds because the serving time was limited. Bey and I always filled our plates and had full bellies after each meal. We were allowed to take food to go so we considered it a blessing. I thanked the Lord for my cargo pants, which were useful and allowed us to satisfy our hunger between meals without driving back to the DFAC during work hours.

The beautiful arch had three windows that I wished I could explore. I was curious as to who lived there: was it the caretakers, the guards, or some of our Special Forces warriors? I asked Bey who might be occupying the arch, and he surmised that it would be the Special Forces. His response made sense. It would offer an effective way to stop terrorists if they attacked. Our warriors could stop the bad guys right before the arch and before they could step onto the SFG compound. It was unlikely to happen, but in a theater of conflict like Iraq, anywhere is dangerous.

The gate to Saddam Hussein's Mother's Palace, Baghdad, Iraq.

DAY AND NIGHT

Bey and I thought that the arch was gigantic compared with the guard towers that we saw in the neighboring camps such as Camp Striker and Camp Slayer. We called it day and night. The barbed wire on top of the T-walls surrounding Camp Slayer was nothing compared with the arch. A guard tower above the barbed wire was a common sight around the camps, except in the SFG compound. The SFG had a double-gated arch and a fortified HQ. After all, they were called Special Forces. I assumed that they might be living under the ground of a spacious marble palace and used the hall for sporting events. After all, like the Tumlin Field sign said, "not all fighting is done outside the wire."

PART III: Back to Iraq. Destination: The Zoo

Guard tower on Camp Slayer.

SFG FANCY DINING FACILITY

Even the DFAC was like day and night compared with the SFG compound and the other camps. Whereas the other DFAC buildings were temporarily built, the SFG DFAC proved to be solid and permanent. The SFG DFAC was very elaborate and gorgeous. It was quite a work of art, and I had a new camera this time, so I took a picture of it. The chandelier looked like a small flower at the center of a giant flower in subtle colors. It might not have been as ostentatious as the Al-Faw Palace, and it lacked Al-Faw's marble halls, but it looked serene. The pillars were decorated with white fonts. There was another entry for a smaller dining section behind the walls for a more private conversation. The SFG DFAC's capacity was not as large as the other DFACs in Baghdad, but it was probably the only building that was adorned with exemplary architecture. The food line was located on the left-hand side of the picture, and there was a small TV in the center of the wall, where viewers could watch Fox News or CNN. There was nothing else to watch but the news. However, the hadji stores had a variety of movie selections for home entertainment.

Special Forces Group's elaborate DFAC.

Bootleg DVDs

While still at the SFG DFAC, Bey and I talked about visiting the hadji stores selling first-run movies, concerts, and other events, instead of watching news all the time. We both agreed that it was challenging for troops and civilian personnel to get access to Hollywood movies legitimately, so they leveraged the abundance of bootleg DVDs in the local shops inside the camps. To ban the troops and civilians from going into hadji stores that sold bootleg DVDs would be detrimental to Iraqi entrepreneurs. Additionally, the coalition forces did not have jurisdiction over shops selling bootlegs. About 20 minutes from the SFG compound was Camp Slayer, which was infamous for a chain of shops selling bootleg CDs and DVDs, as well as different souvenirs from Iraq, such as belly dancing costumes, scarves, women's clothing, leggings, men's clothing, fake watches, sunglasses, belts, mobile phones, water pipes, jewelry, wallets, handbags, table lamps, small furniture, carpets, and many other things that troops and contractors haggled over to get a better price. Everything was subject to negotiation, and often, the hadji yielded to our prices in order to sell his merchandise before the end of the day.

The practice of haggling over prices occurs in markets and stores around the world, including the Philippines. Oftentimes, buyers do not walk away empty-handed. I told Bey that I had watched my mother several times haggling over items and beautiful clothing for me. The seller often chased her and sold the items at my mother's price, which was less than the tag price. It was an amazing technique that I grew up with, only to lose it when I came to America. Bey asked me with sarcasm why I blamed America. I answered that, in the United States, bargaining over a purchase is often confined to yard sales and flea markets, with occasional attempts when purchasing houses and cars. Haggling for any potential purchase will not only give greater value; it can also make shopping a lot more fun. Bey agreed, and I admitted to him that I found it fascinating to shop at the hadji stores. Once again, I could use the haggling skills that I learned from my mother, which always worked. Bey said he liked this skill and could use my help next time we went to the shops.

With my haggling skills, Bey and I were able to buy a collection of James Bond movies, a Star Wars collection, the movie *50 First Dates*, and many other pirated deals. I brought some of these collections as gifts, only to find out later about the prohibition against bringing them back to the United States. The rumor that we could take home a couple of bootleg copies was untrue. Troops and civilians leaving Iraq were subject to customary customs inspections, whether shipping footlockers through the post office or using military shipping MILVANs to send their belongings to the United States. The inspectors were looking for weapons, murals, artifacts, and anything that might be considered an illegal spoil of war. Basically, they were looking for anything beyond flags. However, they did not expect to find any of these items, so they spent most of their attention instead on the bootleg DVD and CD collections. I hoped that the troops and civilian contractors enjoyed their bootleg movies before leaving Iraq, because these items would not make it to the United States. They were prohibited items, and the post office workers conducted white-glove inspections and rummaged through the boxes and footlockers being shipped, as some people tried to find a way to ship them. What was made in Iraq had to stay in Iraq regarding the bootleg copies.

Big Bird at the SFG DFAC

Bey and I had been having an intense discussion about the big bird ever since we saw it. I said it was an eagle with nice, colorful feathers. I thought that the design was unique, but far from the Screaming Eagles vector logo. The Screaming Eagles is the nickname of the 101st Airborne Division, which originated from their insignia, a bald eagle on a black shield. The eagle on the patch is called Old Abe in honor of President Abraham Lincoln and was originally the mascot of a Wisconsin regiment during the Civil War. But according to Bey, it was not an eagle. It was Elmo, the big bird. He was really funny, and we could use such humor—but the big bird did truly resemble an eagle. Bey and I agreed that, in the other DFACs, we did not feel the same tranquility as we felt in the SFG DFAC. There was less commotion in the SFG DFAC than at Camp Slayer, Camp Striker, Camp Cropper, and Camp Liberty. I told Bey that the DFAC exemplified the SFG itself.

Known as the quiet professionals, the SFG consists of the most elite warfighters in the world. They silently slip into hostile countries to train guerilla forces. They know about every weapon in the world, and their medics can take lives as quickly as saving them. They have professional skills in modern war. They allow their sniper teams to work on their skills in a stressful and realistic urban shooting environment. I had not seen women in their unit, but I had heard about the one and only woman to pass the Special Forces Qualification Course so far, Captain Kathleen Wilder, who was assigned for two years to the 5th SFG. I told Bey that I could have been the second female SFG warrior, and at those words, he almost choked on his hamburger. After drinking a glass of water, he said he respected my ambition, there was nothing wrong with it, and that I should have given it a try. I wanted to work in their personnel brigade, but I landed in the 101st Airborne Division's PSB instead. Bey said that maybe it was not meant to be, and God had other plans for me. It was really nice of him to say that. If he had not mentioned God, I would have said that he was just full of it and playing games on me. I also told Bey about a training ground specifically for the SFG at Fort Campbell, and he said I knew too much, and I should keep it with my silent freedom. I smiled and told him that was the perfect thing to do.

There was not much traffic of soldiers and civilian personnel in SFG. It was quiet, and it felt like we were in a different world, a very secure world surrounded by our fierce, lethal Special Forces Group. The SFG compound was exclusive to the Special Forces warriors, and they were a small group. They were the 5th SFG assigned to Fort Campbell, Kentucky, and specialized in operations in the Middle East, Persian Gulf, Central Asia, and the Horn of Africa. The 5th SFG and two of its battalions spend roughly six of every 12 months deployed to Iraq as part of Combined Joint Special Operations Task Force-Arabian Peninsula. They were the first ones I met out of Fort Campbell during this deployment. I had also been to their home base at Fort Campbell when I was with the division's PSB. We conducted personnel records reviews and evaluations, processed promotion packets, and managed personnel readiness records. All the SF warriors I encountered were professionals and gentlemen. They are special indeed!

The author in front of the big bird inside the SFG DFAC.

PART III: Back to Iraq. Destination: The Zoo

FAREWELL PARTY

Before the Halloween party in 2007, Bey said that he would be going home to the United States for Christmas and would not be returning to Iraq. This did not come as a surprise, as he had been talking about it, but he had never mentioned that he was not coming back after Christmas. I felt sad, but instead of feeling depressed, I put on a happy face and planned a farewell party. Luis, Rosemary, and Amanda, the other contractor in HR, also understood the sentiments, and despite feeling crushed, we all got together and planned a farewell dinner at Camp Striker. We also bought a memorable souvenir watch for him from Iraq. He appreciated it and said that he would wear the watch for life. Liar, I said, and we all laughed. It took us an hour to celebrate the farewell party and wish him the best of luck. I thanked him for all the driving lessons, and Rosemary said that I owed him a lot. Because of him, I could go places in Iraq. That was Rosemary's joke. We were forbidden from going outside the wire in Iraq, and we were limited to the luxury of the dining facility, the chapel, shoppette, mail pickup point, and other facilities inside the camp. External boundaries were strictly forbidden. But even then, having the ability to drive outside the zoo was a luxury, and it helped maintain my sanity. Being in a lockdown for a year or more because of a hostile environment is not healthy. It was great that the U.S. government allowed us to use the facilities within the wire. This allowed me to go to church for Mass, use other facilities, and see different activities in the camp.

SAD NEWS

On January 2, 2008, we heard sad breaking news. SSG Ryan Maseth of the SFG had died in Baghdad of injuries sustained in a non-combat-related incident. He died from an accidental electric shock. I remembered seeing him a couple of times when Imee and I used the gym. Some of our fellow employees were friends with Ryan. It was indeed a sad day. He was a brave soldier and went outside the wire almost every day, and it was shock to us to know that he had died of a faulty electrical wire. This reminded me of an incident at Camp Udairi, Iraq. I had almost become a non-combatant casualty too because of a close encounter with two dozen dogs when I was doing my morning cardio exercise. I was running when a large group of Iraqi dogs chased and surrounded me as if they had human brains. Their loud barks fortunately alerted my fellow soldiers, and SSG Figueroa rescued me. Ryan's death was a hard blow for all of us at the zoo, and we attended his memorial service inside the SFG compound, outside their headquarters. It was one of the best memorial services I have witnessed on the battlefield. The pastor said opening prayers, and Ryan's fellow SFG comrades gave their eulogies. We heard about Ryan's life in the field and about his good sense of humor. After the eulogies, a missing man formation aerial salute was performed in memory of Ryan. It was a wonderful salute to a great man.

The faulty electrical wire impacted most of us at the zoo. Since the incident, we were told to think twice when turning on the shower or anything in the bathroom. We also had to use a dry cloth to turn on the water and do the same thing when turning it off. It was like this for a while until we were told that it was safe. The engineers assured us that the wires were not faulty. After the incident, most of us were not sure whether we could trust the engineers in Iraq anymore. After several months, though, I learned to live with it and started turning on the water with my bare hands. It was slow at first, and then I got accustomed to it. When I redeployed and started working in D.C., I stared at the faucet first and hesitated before turning on the water. Even at home, I had to think twice before turning on the water in the bathroom and in the kitchen. One day, I jokingly asked our building engineer if the wires were good. He wondered why I had asked, and because he was a fellow Veteran, I told him about the faulty wiring. He thought I was still bothered by the incident that had occurred such a long time ago, and I admitted that I was. He assured me that all of the wires were grounded and were not faulty, so I should quit worrying so much. It was reassuring, and I trusted his words. But the memories still linger.

The Snowfall

On January 11, 2008, we woke to the novelty of white flakes drifting to the ground, as Baghdad experienced its first snowfall in more than 100 years. The snow melted as it hit the ground, but the news stated that it lingered on the mountains in the north and east, where it was more common. I wished I could see the Tigris and Euphrates rivers covered with snow. I was quite sure that it would be a delightful sight. I was careful when I stepped outside so as not to fall. I enjoyed the glittering snowflakes, which looked like carpets on the ground. I wished that it would last for at least a week so we could enjoy peace and the cool temperatures.

The snowfall lasted for only a day and caused a halt in the gunfighting. There were no sounds of the battlefield: no blasts from explosions, no air-raid sirens, no bombs, no gunfights. The silence was deafening on that day, not hearing a single shot. No one was daring enough to fire a shot under the snowfall, and everyone must have been enjoying the freezing temperatures under their blankets and appreciating the flurries drift by their windows. I was wondering if the snowfall was like manna from Heaven, causing peace on the battlefield. For the Iraqis, it meant hope that would purify their hearts. The news further reported that the streets of the capital were largely empty as big, thick, wet flakes fell. The temperature hovered around freezing, and the snow mostly melted into gray puddles when it hit the ground. Because they melted so quickly, the flakes failed to provide a covering of snow. Nevertheless, many local people still had positive things to say about the once-in-a-lifetime event. It was still lovely, and such snowfall brought a great deal of pleasure to all of us at the zoo and to the people of Iraq.

The snowfall was significant on the battlefield. It seemed to cool the insurgents' heads and stop the war, even if only for one day. It was thought by the Iraqis to be the first snowfall in living memory. The news showed children playing in the snow, while men and women made snowballs in their hands as the snow covered the windshields of their cars. I told Imee that perhaps it was what the people needed in this war-torn country so the fighting would stop. On that day, I wished that it would snow every day in Iraq, Afghanistan, and all the places where our armed forces are in harm's way. The snowfall scared the enemy so they just stayed in place. With my silent freedom, I wished that it would snow for at least 10 months so there would be peace in Iraq.

R&R Time

It was R&R time, and Luis took a two-week vacation in California. Even though he was gone for only two weeks, it seemed like forever to us. Rosemary took charge, and we continued the business of taking care of our linguists and the site managers who came to the zoo for their status reports. The site managers were also taking a big risk by deploying to outlying areas. The fear of the unknown was always there, and no one knew what would come next. Some of the site managers took ground convoys, whereas others

PART III: Back to Iraq. Destination: The Zoo

traveled by helicopter. They always came to the HR office to submit their reports and to talk to the company director in Iraq. There might be more than just a report when they needed to talk directly to the director. It might be something serious regarding changes in military strategy or concerning one of the linguists. Being in HR, we had to know everything because the paperwork trail would come through us.

The surge of troops and civilian contractors increased the traffic flow in Iraq. I recalled President Bush's speech from 2007, when he said that the surge of troops would not yield an immediate end to suicide bombings, assassinations, or IED attacks. Yet, over time, we could expect to see Iraqi troops chasing down murderers, fewer brazen acts of terror, and growing trust and cooperation from Baghdad's residents. When this happened, daily life would improve significantly, and Iraqis would gain confidence in their leaders. The government would then have the breathing space it needed to make progress in other critical areas. I also remembered the fact that most of Iraq's Sunni and Shia wanted to live together in peace. The hope was that the surge would reduce the violence in Baghdad and help make the reconciliation come to fruition.

One of the commitments to delivering a better life was that the Iraqi government would spend $10 billion of its own money to reconstruct and create infrastructure projects that would provide new jobs for the Iraqis. I thought that this was a smart move, as it would help the Iraqi economy and reduce the dependance on collecting bounties for killing Americans. The funding would help the Iraqi people further their entrepreneurial skills and their architectural abilities, thus helping their families and their country. There were rumors of building American retail stores in Baghdad as part of the agenda. The rebuilding would also help empower local leaders, support the Iraqis' plan to hold provincial elections, and allow more Iraqis to re-enter their nation's political life.

Three Musketeers

We made the most of Luis's absence at work while he was on R&R. But for me, things had not been the same since Bey had left HR in Iraq. Although I had the liberty to drive the stick-shift vehicle, it was not as fun without Bey smirking in the passenger seat and criticizing my ability to drive. However, I met Imee, also a Filipina, who had just come on board at HR, and then Lydia, who was also a Filipina. We became acquainted and became gym buddies. Imee and I frequented Camp Striker and VBC. We used their gyms, showers, and dining facilities. I remember Imee's favorite exercises: She liked using the bench for sit-ups and crunches and the exercise ball, turning to the left side and then to the right. My favorite was the total gym equipment. Circuit training burns more calories and increases metabolism. It increases lean muscle mass throughout the entire body, whereas most modes of cardio training primarily involve leg muscles. I remembered Bey's words from when I had worked out with him. He said that higher levels of lean muscle mass equate to a higher resting metabolism, which means that the body will burn more calories while at rest. Adding five to seven pounds of lean muscle mass can increase metabolism up to 50 calories a day or 350 calories over one week.

I found this to be true when I bought a total gym for my house, a few years after redeployment from Iraq. I needed to watch more carefully, though, as I gained more muscles in my arms if I did it every day. After a few weeks of working out, Imee noticed a curve in her waistline. She was full of excitement and told me that waking up early in the morning and driving to the gym to work out had paid off. I asked if she felt more muscle tone, and she said that it was more than that. She said that she was losing weight and could wear the old wardrobe she brought with her. Now, I could see the incentive of working out: to fit in that old clothing.

She laughed and said that we would get fat if we just sat in Iraq without working out. We actually needed to work out for more endurance in our jobs and in doing mail runs. Imee liked my logic and suggested that we encourage Lydia to work out with us.

Doug, a fellow employee, called us the Three Filipina Musketeers, and I thought that it was funny but fitting. He said we were just a perfect group. God is good. Bey left, and God replaced him with two nice walking buddies. I was happy to have the two ladies, especially when we could speak the same language. We found it strange that we met in a strange place, on the battlefield of Iraq. What made it better was that all three of us were Veterans. Imee had served for four years in the U.S. Army. Lydia had retired after 20 years, also in the Army. She said that was long enough. I had also retired from the Army. Imee became my constant walking buddy because Lydia liked to sleep in. Imee and I were early morning risers, and we frequented the SFG compound, Camp Slayer, Oasis, and Camp Striker to use their gyms, shower facilities, and DFAC. We took turns driving the stick shift. It was a wonderful life—except that we were not in paradise. We were in a battle zone.

We Three Musketeers were hard workers, and we all worked in HR for about 16 hours a day taking care of the linguists' benefits, emergency leaves, and many other administrative issues pertaining to job and family. Doing it every day was stressful, and stress has an impact on wellness, so we decided to go to the gym and exercise. This helped a lot, as it refreshed our minds and souls and gave us more energy to work again the next day. Before Lydia came on board after Christmas, Imee and I went to Oasis a lot, especially during the Christmas celebration. There was a long wait in the line of soldiers and civilians trying to get steak and lobster for lunch. I vividly remember all of the soldiers and civilian contractors enjoying their meal together like one big family. Except that the only one who cooked was the chef, who had all the secrets of good cooking. On that day, there was nowhere else I wanted to eat but Oasis.

The food was great, and I enjoyed Imee's geniality. She was daring and often knew how to quickly make her way through the food line without upsetting anyone. For example, she would say that we were in a hurry, as we needed to get back to work immediately, and she would start telling jokes, which entertained the people in line and let us get through and be first in line. But it was true: We were always in a hurry to get back to work. Imee told everyone goodbye once we got our food, and sometimes, she would blow a goodbye kiss. We all laughed at her good-natured personality. There was no one else like Imee. Her friendliness cheered everyone at work and cheered others, too. She also liked helping others, even when it meant fixing roads. One day, Imee and Lydia went to pick up the mail from the hangar. We rotated the mail pickup, and it was their turn to do the mail run that day. When they returned, they were telling everyone about a pothole that had never been fixed and was a threat to vehicles. They returned to the pothole with tools and scooped dirt and rocks from the roadside and filled the hole. Lydia said Imee was really on fire and swept all the dirt into the pothole until it was drivable. The passersby applauded and told them, "Good job!" Imee and Lydia did a curtsy to say thank you. Imee was such a joy, and there was never a dull moment with her. But I discovered later that Lydia was even more alive than Imee. She was always happy and forgiving. She got along with everyone at the zoo, including the gate guards from Fiji. She also got along with the chefs and TCNs at the DFACs. I wished that she was an early riser like Imee, so we could all walk together and start the day with cheer.

It was Lydia who noticed the date palm tree next to our worksite. One day, during our break time at work, the Three Musketeers took a ladder up to the trees and picked dates. Some of the dates were ripe, as they were soft and brown. The others were still yellow, so we decided to give them another week and maybe they would be ripe by then. We showed all the dates to Delphine and to everyone in the office. There were cheers, and one of the U.S. linguists said that dates typically begin to bear fruit in April or May and are ripe for the picking around late August to September. One tree can produce about 200 to 300 pounds of fruit per year, and it can be harvested many times throughout the season, as dates do not all ripen at the same time. This was good to know, as we were in the month of August, a perfect time to pick dates. We also brought some to Doug and his team at the fleet. They were delighted to see our harvest and said they were looking forward to receiving more. Imee said that we would show them where the trees were so that they could pick their own dates. But Mr. Arshad said that the dates would taste better if they were picked by women. Lydia said that the men were not liberated from picking their own dates. With that, everyone laughed.

Luis Returns from R&R

Luis finally returned from R&R, and we welcomed him with lots of surprises. The project manager posted a little sign on his office door that said "FOR RENT," and we planned to explain to him that it had been rented to the highest bidder. There had been multiple bids, but the successful bidder was Subway. Several days earlier, Rosemary had picked up a lot of napkins from Subway at Camp Striker, as well as more napkins from the other stores. Then, the night before Luis's return to the zoo, Rosemary covered the whole front of the building with napkins from Subway and other stores. The resident project manager finished the joke by hammering a sign on the door that said "FOR RENT."

Finally, Luis showed up in the afternoon with a big grin. He looked happy and refreshed. We noticed his new beard, and he looked like a movie star from Hollywood with his sunglasses on. Everyone posed for a picture with him. The project manager stalled him on the road so that we could put the finishing touches on the exterior of his office. After giving him the signal, the project manager walked Luis to his office. He was surprised to see the mess. Several Subway napkins were posted all over his door and even on his windows. It was a wild gag, and Luis's guffaw was heard all over the zoo. We took souvenir photos for him to send to the HQ so he could request a new office. He gave another guffaw, and everybody at the zoo heard it again. It was a great welcome for Luis, and he said that he better not do another R&R so that we would not rent his office. It was our turn to laugh, as we told him that we truly enjoyed messing up his office exterior.

Luis said it was hard to stand up after sitting on the long return journey. He was glad to be back and to see the smiling faces at the zoo. After flashing his genuine smile, he said that he had fulfilled his promise to come back and bring good weather to Iraq. He had had a good time with the family in California and Guam, and he especially enjoyed the sunny weather. We knew that the part of bringing sunny weather was a joke. Who could beat Iraq's sunny weather? We all chuckled at his joke as he continued telling us about his adventures. Two weeks later, it was Rosemary's turn to go on R&R. Destination: Thailand. Bey and I drove her to the airfield and stayed with her until it was time to board the airplane. Little did we know that she would not be coming back to us.

PART III: Back to Iraq. Destination: The Zoo

Rosemary's Disappearance

Three weeks later, it was time for Rosemary to return from R&R. However, she did not show up to work. We got worried and asked Luis about her status. He called Rosemary's sister, who said that Rosemary had not shown up at home either and that all they knew was that she was in Iraq. Rosemary was missing. Luis told us that it had been reported to HQ and that they had hired someone to investigate her disappearance. One of the site managers said that he had seen her in Baghdad. We wondered whether she could be working for another company now, but it was confirmed that she had not tendered her resignation or filed for an extended leave.

The rumor that someone had seen her in Baghdad might have been a hoax as no one else had reported seeing her. If it was true, then at least we would know that she was alive and had made it back to Baghdad. Her disappearance was alarming, and Luis did everything to find her. Two months passed, and we had seen no sign of Rosemary. It was the talk in the zoo for a while until her replacement came on board. Delphine assumed her critical role as the supervisor while Rosemary's investigation was taking place. We welcomed Delphine wholeheartedly and were ready for the changes in the office. One of the changes was to move from Z-1 to a bigger building that could accommodate the entire HR staff, as well as finance, project manager, and many other desk positions. Bey left the zoo and repatriated to the United States before the changes took place. I know that he could have handled the changes because I know how he easily adapts to new environments. But he said that he had to go because his family needed him. He had mentioned to me earlier about a job in Afghanistan, though, saying that it was possible that our paths might cross again. I did not say amen.

New HR Supervisor

Delphine took charge with ease. She got along with everyone. Taking over the HR department is a formidable challenge for any new HR manager, whether newly hired or promoted from within. Aside from functional expertise, a new HR manager's repertoire should include business acumen, a reputation for handling confidential matters with discretion, and accomplishments that indicate an ability to be an advocate for both the company and its employees. An HR manager must know how to influence change, mostly in attitudes about the return on investment in our activities. Such an investment on the battlefield is to maximize productivity by training and taking care of our terps and help our troops in working with local communities on civil affairs projects and other critical missions requiring their service.

After meeting with everyone, Delphine said that she was looking forward to implementing change and that she believed in working together. It was not her first job overseas, so she was used to the heat stress and would help in developing a human capital strategy, as it was her forte. Days passed, and we found out how easygoing Delphine was. She believed in genuine camaraderie and mutual trust, as we were in the mission together in the battlefield. She mentioned that she had two children and that one was finishing a medical degree. As days passed at the zoo, we enjoyed the social gatherings and Korean barbeques that Delphine prepared. We celebrated birthdays, the Fourth of July, Labor Day, and other holidays. We got meat from AAFES, and we ordered catered foods. We invited site managers and linguists who happened to be at the zoo taking care of business.

Delphine was a good supervisor. She instilled mutual respect among us, and she believed that each one of us had valuable and important contributions to make to a successful working relationship. She believed that we were hired by the company for a reason, that we had talent, and that the company valued us for what we were and what we brought to the table. She said that the company saw our unique contributions, recognized and understood our differences, and celebrated our diversified group, while also capitalizing on our common mission, which was supporting the warfighters in Iraq. I thought that her words were well spoken and motivational.

I learned to cook Korean barbeque from Delphine. It seemed that I was learning so many new things on the battlefield, and Delphine giggled when I said that. First, I learned to drive stick shift from Bey, and then, I was learning to cook Korean barbeque. She said they went hand in hand. First, I needed to drive the stick shift before I could buy the ingredients for Korean barbeque from the PX. We both laughed again. She said that it was really easy to cook the barbeque and gave me a list of ingredients: ½ cup of soy sauce, 2 tablespoons of water, 1½ tablespoons of brown sugar, 1 tablespoon of minced garlic, 1 teaspoon of Asian sesame oil, 1 teaspoon of grated peeled fresh ginger, ½ teaspoon of freshly ground black pepper, and ½ cup of chopped scallions. I was curious where she got her ingredients, and she said with a big smile that she had had them since she had arrived in Kuwait. She amazed me, and we were both happy as she continued with her recipe. Whisk the soy sauce, water, brown sugar, garlic, sesame oil, ginger, and ground pepper in a medium bowl. Then, stir in the scallions. Cover and chill up to one day prior to cooking. I still remember her recipe to this day.

The Company Spirit

Imee, Lydia, and I volunteered to do the mail run every day, because Luis said it fell under HR's functions. Delphine was very happy when we volunteered to do the mail run. It was an additional duty on top of what we usually did at HR.

We said that we could do it all the way and that there was no need to appoint anyone else. What a wisecracking trio. We alternated every week, and we simply got a kick out of it. We were upbeat, and we even sang on our mail runs because the vehicle did not have a radio. Before we reached the hangar, one of us got out to be the ground guide. We followed the coalition force's rules of risk management addressing the root cause of accidents. The ground guide should walk in front of the vehicle and to the left of the left fender, to observe traffic to the front and rear of the vehicle.

Risk management is an important way to avoid losses in terms of deaths, injuries, or damaged equipment, anywhere and on everything that is done on the battlefield. Ground guides are a vehicle operator's "eyes" when maneuvering equipment in areas of limited visibility. The ground guide should immediately signal the driver to stop if there is oncoming traffic to prevent collision. A ground guide should wear a safety vest or safety belt for visibility, especially when it is hazy or after dark, and use a radio for communication. Fortunately, we did not pick up the mail at night, only during daytime. Nevertheless, a safety vest was still mandatory. However, I agreed with the importance of the correct positioning of the ground guide. I remember an accident at Fort Campbell, Kentucky, from after my redeployment. One of the soldiers was pinned to the wall while serving as a rear ground guide. A break in communication and the incorrect placement of the rear ground guide resulted in the driver backing up and accidentally pinning the ground guide, causing asphyxiation.

I happened to know that soldier. She was one of my favorite soldiers in her battalion, as she came by our office daily to submit reports or when she needed something from the S-1 shop. She was a good person, had a good attitude, and was always jolly. She was also hardworking. She was not afraid to travel in the combat zone and was always cheerful in helping others. She was probably in her mid-20s when she deployed and had survived serving in the combat zone only to die after redeployment to the United States. It was a tragic accident and could have been prevented if the driver had looked to the rear and ensured that she was visible. Just continually backing up and forgetting about the ground guide was a tragic mistake. It was a very sad day for the 101st CAB. Her death motivated the leadership to spend extra effort to define risk assessment further and establish more rigid rules for ground guide risk management. It was too late to save her life, but the modified rules would prevent similar tragedies in the future.

Mail Runs

As I focused my attention on the road, I was reminded of the importance of a ground guide. Having a vehicle guide was the rule, and if we did not have one, they would send us back with no mail. When it was my shift with Lydia, she exited the car and started bouncing along the stony road, or sometimes, she hobbled or limped to make me laugh. She guided the vehicle up to the designated parking space, and then, we checked in with the postmaster, who authorized us to go to our designated mailbox and collect our mail. Oftentimes, we got heavy packages for our people and had to use a dolly to make it easier to load and unload the trunk. We felt like Santa Claus picking up these packages, especially during Christmastime, and we loved to see the happiness it brought to our fellow workers at the zoo. On our way back, Lydia and I took turns, and I guided the vehicle up to the designated free zone. I pretended to be limping or prancing to keep Lydia entertained while she drove the vehicle. Sometimes, we hollered at people, who looked at us and waved. Lydia and I always got a kick out of this. Even the gate guards let us through easily.

Lydia after she had exited the car to be the ground guide up to the hangar where we picked up the mail.

Lydia and I thought that they might have thought we were both crazy or that we were just having too much fun. Delphine said we were the spirit of the company. She said that it was always fun at the office because of us, and we rose to the bait to try to make it funnier. We were comedians. Often, Imee, Lydia, and I went to the DFAC, and Delphine would tell us to remember to come back. We always did come back, of course, and she said that she was just trying to be funny. She was becoming a comedian too, and she knew how the Three Musketeers could become engaged in an action and might forget to come back to the zoo. We all laughed at Delphine's joke, and we responded that we were sure that we would be coming back. However, no matter what, we would always come back to the zoo with lots of mail, safe and sound. Then again, we might not have because of Imee's curiosity.

Yellow Zone

One weekend, Imee, Lydia, and I had lunch at Camp Slayer. It had become our favorite spot because it offered equally good food and more facilities. Camp Slayer also offered multiple hadji stores and a small post office that we used to ship packages to our families and friends. We heard about a store selling cargo pants and battle shirts. I thought I needed a couple pairs of pants, so I bought some. More cargo pants for the mess hall, I said silently. I did not buy a new pair of boots. I would not trade my boots for anything else. I heard that these were the type of boots that the Air Force warriors wear. I tried them on, and I learned to like them much better than my old combat boots. They felt wide and more comfortable to wear, perfect for my deformed foot. My deformed foot limits me from running. Before my third deployment to Iraq, I went to talk to a doctor. We had a misunderstanding about having surgery on my foot. He mistakenly scheduled me on the wrong day. I thought that the surgery was not meant to be and that I would suck up the pain and have it when I got back from Iraq.

We followed a mine-resistant armor-plated (MRAP) vehicle on our way to Camp Slayer. It belonged to the U.S. Marine Corps, as shown on the back of the MRAP. It also read, "When signaled, proceed and pass with caution," both in English and translated into Arabic. We literally followed it and took these instructions to heart. We would not pass MRAPs unless we were signaled to proceed and pass them with caution. The MRAP reminded me of Patrick's incident when he was serving as a U.S. Army captain. His MRAP ran over an IED. IEDs caused the majority of deaths among our troops in Iraq. I thanked my Creator for saving Pat in that incident. He said that they just had a flat tire and that the hardened MRAP saved them from a destructive device meant to destroy, incapacitate, harass, and distract our troops from accomplishing their missions. MRAPs are built higher from the ground than Humvees. They use big tires that are about 3.5 feet in diameter and cost about $600 each. These tires are designed to run flat for up to 50 miles if necessary. Similar tires are now built for commercial cars. I know this because my car has tires that can run flat for up to 50 miles and are expensive, at about $600 each tire. The MRAP has a warfighter or gunner in the turret.

In a later conversation with a warrior, I learned that a standard gunner had to have an Objective Gunner Protection Kit, or OGPK. An OGPK is lifesaving, standard-issue equipment for tactical Army gunner vehicles, from up-armored Humvees in Operation New Dawn in Iraq to OEF in Afghanistan, in addition to MRAP vehicles. An OGPK is an integrated armor ballistic glass turret that is mounted onto the turret ring of tactical and armored vehicles. The kit also has a turret shield, gun shield, and all necessary hardware for mounting the system to the vehicle. Army teams were responsible for developing this lifesaving product, and they continue to develop more lifesaving devices that will save many troops and civilians in the battlefield, such as Iraq and Afghanistan. My silent freedom screams that if we continue to engage in war, we might as well bring the most lethal weapon that will stop the enemy, so that we can continue to accomplish our missions efficiently and effectively. I wish that I had the gift to invent such tools. That is why I thank God for my silent freedom; I can think what I want to think, and dream what I want to dream.

MRAP owned by the U.S. Marines.

As we received the signal to proceed and pass the MRAP, Imee hit the pedal slowly, and with caution, she drove past the MRAP. We waved at the warfighters in the vehicle, and they waved back at us. I thanked God for our brave warfighters. Lydia and I did not know where Imee was taking us, but she had suggested that we eat authentic Iraqi food, and Lydia and I just went along with the idea. We stopped at a small restaurant, and for me, it was not really different from what I had had in Dahuk, Iraq, during the first deployment in support of OIF. Nevertheless, I agreed with the other ladies that the food was good. Imee had another idea to go to BIAP, which was about a minute's drive from the restaurant. Lydia and I both looked at Imee asking if she was serious about the idea. Imee said that she was a Taurus and therefore stubborn. We laughed when she said that, and she said, "Come on! Let's go!" We used the same car, and we could not allow Imee to drive by herself to the airport just to quench her curiosity. We left the restaurant and hopped into the car, with Imee as the driver. The Iraqi restaurant owner had been right; the airport was not far at all.

We all got out of the car as soon as we reached BIAP. Imee did not do combat parking, and we all rushed into the airport to check it out. It was not bad at all, and we saw that every woman was wearing a niqab, a veil for the face that leaves the area around the eyes clear, with an accompanying headscarf, and an abaya, a long black cloak worn as outerwear over a dress or pants. They all looked at the three of us as we entered. I did not trust their stares. After two minutes, I urged my friends that we needed to get out of there as soon as possible. Imee wanted to explore a little bit more, but I persuaded her to leave as fast as we could. We all got into our car and drove rapidly away from the airport. Imee exited safely with no incident. Imee's curiosity could sometimes put us into trouble, and my throbbing headache disappeared as soon as we passed the little restaurant where we had eaten authentic Iraqi food. Nevertheless, it was an experience that will be remembered for a lifetime. We found out later that the area where we ate and the airport itself were designated as the yellow zone and were forbidden to our American soldiers and contractors, although we occupied the other side of it.

Unforeseen Dust Storm

We went through a long line of cars waiting to get through the gate to get back inside the coalition forces camp. It was hot and humid, and we were glad that the car had air conditioning. The gate was secured with barbed wire and more T-walls. The checkpoint was strict, and we all showed our badges, in addition to Imee's wink at the guard. Imee was really funny, and it often worked. Lydia and I just laughed it off. We stopped at Camp Striker to refuel. A man with an accent said that he was from Pakistan. He started spilling out a story about his family and his little girl back in Pakistan. Imee was a live wire and kept him talking until it was our turn to top off the tank. She signed off on the paper and told him we would see him again soon. The man smiled, put his hands together, and thanked us. We noticed that it was becoming hazy, and Imee slowed our vehicle. She was the designated driver, and Lydia and I entrusted our lives to her hands. As shown in the next picture, the weather was getting bad. We did not have modern smartphones at the time that could show the weather forecast in Baghdad.

We were about five minutes away from the zoo. Imee spoke like a queen and told me to start praying. Instead of smiling, I closed my eyes and asked for the Lord's intervention and for us to reach the zoo before the dust storm got worse. Lydia was more vocal with her prayers and kept saying, "Oh God, Oh God, please let us get to the zoo first before it worsens." Dust storms can take down trees, bury equipment, and potentially damage houses and cars with passengers in them. In that situation, we fell in the latter category. The toxic dust affects anyone without a face covering, especially those with pre-existing respiratory conditions. I was not sure whether my two companions had such problems, but at that time, we were focused on driving quickly and safely and getting home as soon as we could. Delphine and Luis were happy to see us when we made it back, and as soon as we had parked the car, we ran inside the building and closed the door. We remained inside for about half an hour and peeked outside once the storm had subsided. The Three Musketeers looked at each other, relieved, and reminisced about the trip to the small Iraqi restaurant and BIAP and the close call of getting caught in the middle of dust storm. It was a risky escapade that I will never forget.

Oncoming dust storm at Camp Striker, Baghdad, Iraq.

We never told anyone about the escapade. I thought that it was dangerous and was leery about it in the first place. It reminded me of my drill sergeant's words: Curiosity can sometimes kill you. You see something in your environment, and before you know it, you are chasing something that could lead you to your death. Those words were alive to me on that day during the risky escapade at the airport. Curiosity might not lead to death under normal circumstances and when we are not at war. However, because of the conflict that was going on, we could have gotten caught in the middle of gunfighting or been held hostage for a ransom. It was a perilous situation, and I swore that I would never yield to it again. But Imee looked as relaxed as usual and did not even display a sign of remorse. She actually looked proud for our short expedition. Lydia, on the other hand, looked guilty like me. She looked like she would burst into tears. I ended up laughing, and Lydia and Imee laughed with me. We agreed that we were going to write a book about our adventure, and we all felt excited.

Halloween in Iraq

It was the year 2008, and it was time for our Halloween celebration in Iraq. I heard later that despite being deployed, most camps had parties that resembled what a typical Halloween party would look like in the United States. At the zoo, we were dressed as several characters, such as clowns and queens, to mention a few of the decent ones. There was going to be a best costume contest, and two months prior to Halloween, my co-workers started ordering their costumes, as did I. Imee asked her husband in the United States to send her a costume, Lydia ordered hers online, and I looked for one online, too. The Three Musketeers decked the worksite with Halloween decorations sent by Imee's husband and other items ordered online. Halloween night came, and everyone showed up in their scary Halloween costumes that would freak out literally everyone at the zoo. To my surprise, no one was dressed as a lion or fox. Those costumes could have been really freaky. I thought that Imee would get the money prize because she was nicely dressed as a queen. Or Lydia, because she was dressed as a pregnant nun. A big face of an old, scary man showed up, and I thought that he would scare everyone, until a skeleton showed up, followed by more creepy participants.

Imee was dressed as Marie Antoinette, and she looked beautiful. She was the epitome of Halloween costumes. She had more than one costume, and earlier that day, she reported to work in a niqab. She was wearing a full traditional black burqa with a voluminous drape. Her headscarf was tied like a bandana. The niqab is a traditional Arab headdress that covers the face and leaves open only the eyes. However,

Imee covered her whole face but could still see through a thin slit in her niqab. Suddenly, the entire staff became excited by her antics and eager to pose for pictures with her. Her costume also shocked the site manager, who was also eager to have a picture with her. We even created a short comedy skit by pretending that we were in-processing her as she went through all the stations. The simulated process was just perfect, until she came to Anita, who belly laughed and then cried. Before lunch, the Three Musketeers went to the fleet maintenance area to show off Imee's new identity. Doug, Tupou, and Mr. Arshad also had belly laughs and cried at Imee's antics. As if it were not enough, Imee spoke a little Arabic, and everyone just went insane. She could be mischievous but was still a talented linguist. She could speak several of the 146 Filipino dialects including the national language of Tagalog, and now she could speak Arabic, too. One day, Doug said that Imee was intellectual and a handful. We both laughed at the description, and Imee did not mind Doug's constructive criticism.

Imee said that the fun was now over and that we were going to eat lunch and do the mail run. It was our turn to go to the hangar and to have fun picking up the mail. She changed into her work clothes so we could go to Camp Striker. We did not let Imee wear her costume to the DFAC, as it might have caused complications that resulted in us getting sent home. As we approached the base, we did not see anyone wearing Halloween costumes yet, but we heard that some units were going to have parties for a few hours and celebrate as most people normally did and loosen up a little bit. We had a good lunch, and then, we rushed to the hangar. It was against policy to stop anywhere after picking up the mail. It was my turn to be the ground guide, so I got out of the vehicle and let Imee drive. As we parked the car, we were surprised to see the mail staff in their Halloween costumes, showing their faces. The man we always talked to was a vampire, and the lady in the office was wearing a witch costume. Others were wearing their regular work clothes and jokingly stated that they looked scary enough and did not need costumes. We all laughed as we confirmed their costumes' suitability and left the hangar cheerfully. We picked up a number of packages. Imee and I agreed that when it rained, it poured. Why was there so much mail when we were in a rush? Nevertheless, we managed to put all of the boxes in the vehicle and saw many smiling faces at the zoo when we returned in a vehicle loaded with packages. We felt apprehensive and imagined how it would be in the few days before Christmas.

Imee in her niqab with the author in front of Z-25, our worksite at the zoo, Baghdad, Iraq, 2008.

At last, the evening came. The Halloween party was finally here, and an assortment of creepy, funny, and beautiful creatures gradually began to arrive. Some were fully costumed and masked; others applied heavy makeup beyond recognition. I picked up Lydia from her billet and was surprised by her costume. She was dressed as a pregnant nun! She laughed when she saw my facial expression. I am Catholic, and I was not expecting to see a pregnant nun! She also made a comment about my costume, and I told her that it was the warmest costume I could bear in 65-degree weather. She laughed and asked why I did not just wear my bomber jacket instead. I told her that that was not a costume, just a jacket, and she burst into laughter. It was so much fun to be with Lydia.

The Halloween costume contest would be announced after the photo-ops and some spooky Halloween dances with the employees' full participation. Delphine finally came out, dressed as a clown, and she started making jokes. We asked if she was going to juggle her spinning balls tonight, but she said that they did not come with her costume. Anita arrived dressed as a Japanese geisha. To complete her costume, she had applied so much thick white makeup that we could not tell it was her. We recognized her only by the way she walked. Later, the fleet team arrived not wearing costumes. They said they looked scary enough and did not need them. It was nice for them to show up, though, and participate in the celebration. I was glad that there was no alcohol, and it was great to see everyone loosen up for a few hours. Everyone was excited to hear who the winner was, so we all stayed until the last minute of the party. The costumes were really good, and I was surprised at the turnout and at how much effort everyone had put into their costumes.

Finally, Luis (the emcee) got on the mike and thanked everyone for their full participation and for trying to make the evening as enjoyable as possible. He said that it was tough to judge the costume competition as all the creative getups looked great. Someone was disguised as Indiana Jones, and it turned out to be one of the bosses. Other costumes included a bat, funny-looking characters, and scary-looking creatures. This night was special, because everyone could wear their favorite costume even in the middle of the desert and on the battlefield. Fortunately, no one was dressed like a tank, or he or she would have run over us. I noticed that we had visitors from the military and the Blackwater company, who might have been invited by the site managers. I was not sure if we had any warfighters from the SFG, because almost everyone was wearing a costume, but some of our site managers also had friends in the SFG. We were glad that there were abundant finger foods and near beers, and everyone had plenty to eat. We even had Halloween sound effects, and the celebration was simply perfect.

Everyone quieted down when the emcee announced the third, second, and then first prizes. The first prize went to the human cadaver, and everyone agreed and cheered. I believe that he won a cash prize and was a happy camper that night. It turned out that the cadaver was our site manager. What an idea! I should have thought about being a mummy instead, and maybe I would have been a prize winner. The event went peacefully and successfully. It was a fun night and a good morale booster. The Halloween party was the first of my holidays at the zoo, so I took some photos as mementos. We might not have another Halloween celebration, but everyone was already looking forward and planning events for Christmas and New Year's Day. Some photos are shown on the following pages to remember the creepy, fun Halloween night at the zoo in 2008. Some of the characters were not as recognizable as they wanted to be and were happy to be called creepy. I was hoping that the troops also enjoyed their Halloween costumes and parties. I had heard that some of the units were planning to play horseshoes and volleyball in addition to having costumes and live music to build morale for everyone. Though the night was fun, it was also long, so we were tired the following day. Everyone was grateful that there had been no incidents during the Halloween party. I can imagine the bad guys' faces seeing us in our creepy costumes. It would have probably been enough to scare them away!

The author poses with the "human cadaver," who won best costume at the Halloween party at the zoo, Baghdad, Iraq, 2008.

PART III: Back to Iraq. Destination: The Zoo

Some creepy and spooky costumes from the Halloween party at the zoo, Baghdad, Iraq, 2008. The author is at the left, and Lydia is at the right.

THANKSGIVING AT THE ZOO

On Thanksgiving Day, 2008, I remember that Kathy, the manager of the linguists, made tepees for Thanksgiving decorations. They were made of patched papers and cut out pictures from magazines. They looked real but not livable, as they could have torn even before we celebrated Thanksgiving. Then, for Christmas decorations, she decided to make snowflake paper cutouts. I helped her with the cutouts and posted them on the windows. They were really

A few more creepy and cute creatures from the Halloween party at the zoo, Baghdad, Iraq, 2008.

beautiful, and the sight made us homesick. What were the chances of getting snowflakes in Iraq again after the snowfall earlier this year? Because there was a high probability that we would not get even a glimpse of snow, Kathy decided to create a paper substitute. Maybe the artificial snowflakes would invite some real snowfall this Christmas. I will never forget how peaceful it was in January 2008 when it snowed in Iraq. It was not enough to build a snowman, but it was enough to stop the fighting, at least for one day. If they stopped fighting for good, then maybe we could leave Iraq, go home to the United States, and enjoy the real snow. By making snowflakes for office decorations, we were wishing that we would get to experience snow in Iraq one more time before we left the country.

Oasis DFAC filled with Thanksgiving decorations, 2008.

Every year, the chefs and DFAC staff warmed not only our stomachs, but also our visions with a well-decorated DFAC. The decorations gave joy to everyone who entered the hall, knowing that they would see something other than dust, storms, and barbed wire. The spectacular Thanksgiving decorations were sent from the United States several months early so they would arrive on time for the event. I remembered my assignment with the 86th CSH as I looked at the decorations. I learned to appreciate the nurses more than ever, as they did similar decorating. They were good at taking care of patients and equally adept at decorating the hospital to commemorate different events and holidays. This was something that I learned and applied to our office in D.C. a few years later after returning from Iraq. I decorated our worksite for every single event, and my co-workers appreciated it and acknowledged the joy that they got from these arrangements.

CHRISTMAS AT THE ZOO

Christmas decorations in HR, Building Z-25, at the zoo, Baghdad, Iraq, 2008.

The Three Musketeers changed the theme from Thanksgiving to Christmas as soon as Thanksgiving was over. I had ordered some items online months before December, and Imee's husband sent us the rest of the decorations from the United States. The worksite had a conference room on the left-hand side, where we celebrated occasions. We had plenty of Christmas cookies throughout the month of December at the zoo. Some of the cookies were sent by families and friends from the United States, and most of them came from Imee's husband. Imee was the queen of philanthropy indeed.

The author standing next to the nine-foot Christmas tree inside Building Z-25 at the zoo, Baghdad, Iraq, 2008.

After having lunch at Oasis, we stopped at the fuel station to get gas. The line was long again, so we had to wait. I became familiar with the workers at the fuel station, and they became familiar with us as well. They called each of us "madame" and started talking to us like their acquaintances. But we were strictly on business and were in a hurry to get back to work. Sooner or later, the three of us would start going to Sather's DFAC to eat. It was my first time to see airmen and -women in full battle rattle uniform. We had always joked that they were not real warriors, but the battlefield changed our impression, as they proved that they could also fight and were part of the fighting team after all. We felt secure under the protection of the U.S. Air Force. Every time we went anywhere, including mail runs, we carried our helmets and Kevlar vests for safety purposes. The piece of equipment I was missing on this tour was my M4. We were not authorized to carry one. If we had been, that would have made my day, because I would have been able to defend myself and my fellow workers from aggressors.

REINDEERS IN THE DFAC

Reindeers in Oasis waiting to be displayed for Christmas.

Even though they were not real, the reindeers in Oasis were delightful to see. They were in the corner waiting to be placed at different stations for Christmas. I asked Doug why we saw a lot of reindeers on Christmas. He said that reindeers live in extremely cold conditions and are known to pull sleds. Then, he added that in a traditional festive legend, Santa Claus had reindeer pull a sleigh through the night sky

PART III: Back to Iraq. Destination: The Zoo

to help him deliver gifts to children on Christmas Eve. Imee chimed in and named the reindeers before Doug could get to it. She said, Blitzen, Comet, Cupid, Dancer, Dasher, Donner, Prancer, Rudolph, and Vixen. Doug asked what was so special about Rudolph. We all chimed in and said that he was added to the original eight reindeers to comply with FAA lighting requirements for aircraft. And we all burst into laughter.

COMMANDING GENERAL'S VISIT TO THE DFAC

We heard that the CG would be visiting the Oasis DFAC, so we stopped by to check. We stopped talking when General Petraeus showed up with his aides and was giving away challenge coins. We hurriedly joined the line of soldiers and civilian workers waiting to have a photo op with the general and receive a coin. General Petraeus is a man of few words. He gave a brief speech and wished everyone a safe and merry Christmas. It was such an unforgettable event. I was surprised he did not recognize me from previous deployments under his command, especially because he had signed both of my Bronze Star awards. Then again, I looked different in my civilian clothes, and he nodded his head when I mentioned that I was assigned with the 101st CAB last year, retired from the 101st Airborne Division, and then decided to return to Iraq as a contractor. He shook my hands and said, "It is good to see you." The CG always said that even when I was assigned to his command. But he connected with troops easily, and we respected him for caring. The CG moved fast too. He left as quickly as he came.

After the CG had left Oasis, we indulged in steak, lobster, shrimp, and scampi. We tried the desserts too until our stomachs were full. There was no rush to get back to the zoo since it was Christmas. We were on call if an emergency arose. We stayed at the DFAC for about half an hour and

Imee receiving the CG's coin, Oasis DFAC, Baghdad, Iraq, 2008.

stopped at the PX on the way back. It was the only leisure activity we had other than the DFAC, aside from the escapade that the three of us had had in the yellow zone. That was indeed an unforgettable event, and Imee actually wanted to visit the restaurant again. I declined the invitation knowing the risk it posed. It was not safe, and we might have been held as hostages or been shot at. Who knows?

The reindeers were finally moved to different stations. The best sight was the two reindeers standing with the Christmas tree, as shown in the picture. Some of my co-workers arrived later and posed for photos with the reindeers. We did not see movie entertainers this time around, so anything that entertained us, such as the cutout reindeers, could generate great enthusiasm. It was such a joyful celebration on the battlefield. The camaraderie was great, and the food was always good. Near champagnes were served, and we all said,

"Cheers!" Lots of finger foods were also served, and our brief opportunity to socialize with the others in the DFAC was delightful. We all enjoyed the food, and I was thankful to finally see the 101st Airborne Division.

Two reindeers placed with the Christmas tree, Oasis DFAC, Baghdad, Iraq, 2008.

It was quite a special night, because I got to talk to the CG and he gave me a challenge coin, which is presented to recognize a special achievement. The challenge coin that I received is a good-sized coin, bearing the symbol of the 101st Screaming Eagles on one side and OIF on the reverse. I collect coins and have several from the Screaming Eagles. That night, the CG thanked us as we continued to fight, support, and win OIF. I also have a challenge coin from the CG's right arm. CSM Hill had been the CG's senior enlisted adviser since 2003, our first deployment to OIF, in Mosul, Iraq. It was a coincidence that General Petraeus and CSM Hill had been deployed again this time around. I worked in Baghdad, and they were located at Camp Victory, in the Al-Faw Palace. It would be challenging, but I dreamed of stopping by to say hello and reminisce about the dangerous experiences of previous deployments as we rendezvoused with destiny. The 101st Airborne Division was known to rendezvous with destiny when it was activated on August 16, 1942, at Camp Claiborne, Louisiana. It had no history, but it had a rendezvous with destiny.

CHRISTMAS MASS AT CAMP STRIKER

On Christmas Day, Doug and I went to the Catholic service at Camp Striker's DFAC. He was Protestant, but he wanted to worship too. He said that the Catholic service was the closest he could get on the battlefield. Imee and Lydia decided not to go to Mass with us. I noticed that the room was filled with civilians and soldiers wanting to worship the Lord on His birthday. The celebration was not as fancy as my previous Christmas celebrations on the battlefield, but the solemnity of the Mass was exemplary. I was thankful for the opportunity to serve; for good friends and camaraderie; for my family; and for the past, present, and future. I volunteered to read the first reading, which was from the book of Titus (2:11–14):

> For the grace of God has appeared, bringing salvation to all, training us to renounce impiety and worldly passions, and in the present age, to live lives that are self-controlled, upright, and godly, while we wait for the blessed hope and the manifestation of the glory of our great God and Savior, Jesus Christ, who gave himself for us that he might redeem us from all iniquity and purify for himself a people of his own who are zealous for good deeds.

PART III: Back to Iraq. Destination: The Zoo

I concluded the reading with, "The Word of the Lord." The people's response was "Thanks be to God." After we had sung the Alleluia in unison, the priest stood up and read the Gospel according to Luke (2:1–14, 15–20):

THE BIRTH OF JESUS

In those days, Caesar Augustus issued a decree that a census should be taken of the entire Roman world. This was the first census that took place while Quirinius was governor of Syria. And everyone went to their town to register. So Joseph also went up from the town of Nazareth in Galilee to Judea, to Bethlehem the town of David, because he belonged to the house and line of David. He went to register with Mary, who was pledged to be married to him and was expecting a child. While they were there, the time came for the baby to be born, and she gave birth to her firstborn, a son. She wrapped him in cloths and placed in a manger, because there was no guest room available for them.

And there were shepherd living out in the fields nearby, keeping watch over their flocks at night. An angel of the Lord appeared to them, and the glory of the Lord shone around them, and they were terrified. But the angel said to them, "Do not be afraid. I bring you good news that will cause great joy for all the people. Today in the town of David a Savior has been born to you; he is the Messiah, the Lord. This will be a sign to you: You will find a baby wrapped in cloths and lying in a manger."

Suddenly a great company of the heavenly host appeared with the angel, praising God and saying,

"Glory to God in the highest heaven, And on earth peace to those on whom his favor rests."

When the angels had left them and gone into heaven, the shepherds said to one another, "Let us go now to Bethlehem and see this thing that has taken place, which the Lord has made known to us." So, they went with haste and found Mary and Joseph, and the child lying in the manger. When they had seen him, they spread the word concerning what had been told them about this child; and all who heard it were amazed at what the shepherds said to them. But Mary treasured all these things and pondered them in her heart. The shepherds returned, glorifying and praising God for all they had heard and seen, just as they had been told.

When the priest had concluded the Gospel, he said, "The Gospel of the Lord." The congregation replied, "Praise be your Name, Lord Jesus Christ." We took our seats and listened to the priest's wonderful homily. To me, it was comforting to listen to the word of God, especially this Christmas, when so many people were away from their homes. The singing and the reading of the scriptures were very solemn, and the fact that we were celebrating our Creator's birthday on the battlefield mattered. He is the reason for the season.

After the Mass, Doug and I browsed the DFAC one more time, deciding whether we were going to take food back to the zoo. Under normal circumstances, if I were to receive the Most Holy Eucharist, I would abstain for at least one hour before Holy Communion from any food and drink, except for only water and medicine. The Holy Eucharist is the sacrament commanded by Jesus Christ for the continual remembrance of His life, His love for us in dying on the cross, and His resurrection until He comes again. Jesus reveals Himself to his church through the breaking of bread in a way that is reminiscent of the time when Jesus broke bread with his disciples and instituted the Lord's Supper.

Again, there are several instances in the Bible when Jesus simply appears among the disciples, frightening them because they think he is a ghost. Jesus reassures them and provides some identifying marker to prove that it is Him, such as the nail marks in his hands or the spear hole in his side. Also, when she was in the garden, Mary thought that Jesus was the gardener until He said her name. Also, when several of His disciples were out fishing, they did not recognize Him until He told them to cast their nets on the right side of the boat, where they caught a miraculous number of fish. Even on the road to Emmaus, two of His disciples were going to the village of Emmaus, about seven miles from Jerusalem, and were deep in solemn and serious discussion, when Jesus met them. They could not recognize Jesus and saw Him as a stranger. Jesus let them share their anxieties and pains and let them grieve and mourn by expressing root causes. Jesus empathetically listened to them, as they cried out about their crises and doubts and used scriptures so they could better understand suffering and glory. The two disciples begged Jesus to stay with them as they reached Emmaus. Jesus then stayed with them, and they did not recognize Jesus until the pivotal moment when Jesus broke the bread (Luke 24:13–35). I believe that Jesus is present at Mass when the priest breaks the bread and during Holy Communion. This also reminds me of the third time that Jesus appears to His disciples after the resurrection: "He took the bread and gave it to them, and did the same with the fish."

My mom read the Bible to me when I was a child, and one bedtime, she read about the Holy Eucharist. The Eucharist is the Church's sacrifice of praise and thanksgiving and is the way in which the sacrifice of Christ is made present and in which He unites us to His one offering of Himself. The Holy Eucharist is called the Lord's Supper and Holy Communion. It is also known as the Divine Liturgy, the Mass, and the Great Offering. The outward and visible sign of the Eucharist is bread and wine, given and received according to Christ's command. On that night, I became immersed in what my mother was reading to me. The inward and spiritual grace in the Holy Communion is the Body and Blood of Christ given to His people and received by faith. My mom continued reading, and slowly she said, the benefits we receive are the forgiveness of our sins, the strengthening of our union with Christ and one another, and the indication of the Heavenly banquet, which is our nourishment in eternal life. My mother instilled in me the discipline of the Holy Eucharist, and I tried very hard as a young girl not to eat anything before going to church and receiving Holy Communion.

My thoughts about abstaining before taking Holy Communion were interrupted when Doug asked what I wanted to eat at eight o'clock at night and if we should take it to go or eat it at the DFAC. I said that we should eat at the DFAC to get it over with and to not to have to clean any mess if we took it back to the zoo. He laughed and agreed that it was a marvelous idea to dine in so as not to attract more mice in the zoo. And he complained that he was running out of mouse traps. We both laughed; however, it was true. We both agreed not to talk about the mice and joined the short chow line. If it was healthy, we would rather go to the DFAC at 8 p.m. when it was not crowded but there were still plenty of selections in the food line. Eating late could cause acid reflux and many stomach problems, especially for Doug. And we talked about a litany of unhealthy habits to live longer. Although Doug was fun to eat with, he was not as funny as Bey. Doug had a dry humor, whereas Bey was a natural, and his wisecracking was endlessly amusing.

After eating, Doug said he had to feed his mouse. I had goose bumps when he said that, and I asked, "What mouse?" He said that this particular mouse had kept coming back, so he decided to feed it more. Doug took two small boxes of dry cereal and some strawberries and grapes. He drove the car, and the subject of our conversation from the DFAC to the zoo was the mouse. He said that mice are opportunistic omnivores, meaning that they generally eat plants, seeds, and grains but they will also eat insects when they get the chance. I asked Doug if the rats were in imminent danger of extinction in Iraq, and he said no. He laughed at my question and said he knew where it was going. We both laughed, and he explained how he found the favorite rat. He noticed that the same rat crawled by his window at a certain time of the night, every night, and gnawed on his pencil. He cleaned his desk and put cereal on it instead. The rat came back and ate the cereal. He said that rats are loyal and affectionate and are very easy to care for. They are among the cleanest creatures that one could find to keep as pets. He added that rats do not shed. I told him that he could continue feeding his purebred rat as long as he kept it away from the HR. I had never heard his guffaw before, and I thought it was funny too: a purebred rat in Baghdad, Iraq.

Doug was the fleet manager, and the Three Musketeers appreciated his generosity every time we asked for a vehicle for the mail run. He would tease us to bring him packages in return, and we would say yes. If he did not have a package, we were going to stop at the PX and buy him one, and he laughed at our joke. He took care of us and supported our benevolence in picking up everyone's mail from the hangar. The mail run was a way out of the office after working for many continuous hours, and having a safe car to drive was a blessing for us and for everyone at the zoo who was expecting mail from the outside world.

Camp Victory Night

Life was zestful at the Camp Victory complex. The Three Musketeers invited Doug for an adventure at a Filipino party, but he declined and asked for a to-go plate instead. The Three Musketeers met some Filipinos who worked at Camp Slayer, some who worked at Camp Striker, and others who worked in other camps. I was surprised that they had been in the country for a while. I used to have neighbors in the Philippines who worked in Iraq so many years ago. I am not sure whether they worked in this part of the country. After exchanging stories, our new Filipino friends invited us to their party. We knew better than to decline the invitation because we were familiar with Filipino traditions. They make sure that there would be enough food for the party and that the place would have enough space to accommodate everyone. They did not mind at all if you invited more friends with you as long as the host was informed ahead of time for food preparation. The concept of Filipino time follows Filipinos wherever they are, even in Baghdad, Iraq. Guests arrive late, and it is still okay. But they can also arrive early to talk and bond with the other early birds or the party hosts themselves before the party starts.

The food being served during the party included lumpia Shanghai, or egg rolls; regular pancit and pancit palabok, or Filipino spaghetti; Kare Kare, which is a thick peanut stew that usually contains oxtail and vegetables; chicken adobo; and sisig, which is made from parts of pig head and chicken liver, usually seasoned with calamansi, onions, and chili peppers. They also had bibingka, or native cakes; white rice cakes; deviled eggs; and many more dishes. The Three Musketeers ate a lot, and on top of that, we brought plates for Doug and to share with others at the zoo. It was an evening well spent at Camp Victory.

PART III: Back to Iraq. Destination: The Zoo

They also had a facility for floor dancing and belly dancing lessons every Saturday night. One night, the Three Musketeers tried the place, and surprisingly, we were not the only ones taking advantage the opportunity to have some fun. There were lots of women in nice clothes, and we wondered where they got them from. The Three Musketeers were wearing their battlefield clothes and were no comparison with the ladies who had just arrived in fancy dresses and high heels. I told Imee and Lydia that there was no way I would wear those heels in the desert. Nevertheless, we took a couple of spins on the dance floor because our favorite music was playing and did a couple rounds of belly dancing. It was always fun to go out with Lydia and Imee.

The night was young, so we explored the other rooms in the facility. Lydia and I found a good corner to pose for a picture. It was holiday season, and I was wearing a tiny Santa hat compared to Lydia's more prominent hat. Imee took our picture, and I took theirs too. After dancing, we decided to go to the PX and feast our eyes on the new mugs and trinkets. I bought a new Air Force jogging set, which I treasured and still have to this day. The mugs mostly came in metal and ceramic and were adorned with the logos of the Army, Air Force, U.S. Marine Corps, and other members of the coalition forces. I particularly liked a mug that read, "Iraq in my rearview mirror."

We stayed in the PX until the last hour and then joined the soldiers and civilians waiting in line for the cashier. My favorite was Helen from Ethiopia. We became friends as days went by, and I always went to her to pay for my purchases. She said that she came from Germany and had been working in the PX for a long time. She applied for the PX in Iraq and got the job. I told her that she was courageous. She sheepishly smiled, and I was impressed by her answer. She said that you should be courageous to get what you want.

If she had not stepped forward, she would have never been part of the change and would have missed the opportunity to help the efforts to improve the Iraqi quality of life. I told her that she was my hero, and we found a trusted ground on that day. God is omnipresent, and Helen said that she was a Christian and believed in the love that God has for us. God is love, and whoever abides in love, abides in God. She was happy to serve the troops and hoped to stay until they closed the shop for good.

Later on, I introduced Doug to Helen. Doug liked going to go to the PX for a mutual reason: It served as a break after working for the whole week. Helen was happy to meet Doug, and from then on, we went to Helen to pay for our purchases. We did not go to the other cashiers. If we had to wait in a long line, then we were happy to do so just so we could chitchat with Helen and pay for our purchases at the same time. Helen was always delighted to see us, and she said that she wished that we could see more of each other. Doug and I laughed and said that that could be arranged. We would do our best to visit her at the PX after the DFAC. Helen was skeptical of Doug's promise, but she believed when we fulfilled our promise to visit her every day at the PX, even if just for beef jerky. Helen's jovial personality boosted our morale and made our days in the field more manageable. We noticed that her fellow workers became more pleasant too because of her example. She had become their role model. When there was a long line to Helen's desk, the others waved their hands at us and helped us with our purchases. Going to Camp Striker became more gratifying because of Helen and her crew.

No One Was More Cheerful than Lydia

Lydia was fun like Imee. There were no dull moments around her. She was always happy and liked to travel the

PART III: Back to Iraq. Destination: The Zoo

Lydia and the author in Camp Victory, Baghdad, Iraq, 2008.

world. She was deployed to Bosnia and came home to retire, just to find herself in the battlefield again. I remember Lydia's excitement when we were in Camp Striker early in 2008. She asked if we could go to lunch with her so she could introduce us to her son, who was currently serving as an officer in the Air Force. Imee and I cheerfully accepted the invitation, and I volunteered to drive. Lydia said that she could not wait to see her son. It was the first time she would see him in the battlefield. Imee and I felt her excitement too, and we never knew that her son was serving in the Air Force until she told us. Sometimes, I felt the same way. I kept things in my silent freedom until the day I could see my son and give him a big hug, and it meant the world to me when he hugged me back. I also felt more ecstatic when I saw them all together: JR, Jodie, Evan, Patrick, Abi, Christopher, and Krishna. They all mean so much to me. Lydia disrupted my train of thought when she asked if we could leave now for lunch. I said, "Of course."

I took the keys from the rack, and after asking Delphine what she would like from the DFAC, she said, "Anything mouthwatering," and we all laughed. I led both Lydia and Imee to the car, and we all suppressed our excitement until the car started and we were on our merry way to Camp Striker. We were all quiet until we reached the gate at the zoo and waved at the guards. They waved back, and we all smiled. Imee broke our silence by telling a funny story, and she kept on and on until we reached Camp Striker. We found a parking space close to the DFAC, and I backed up to park. Lydia said I was getting good at combat parking, and I responded that it was the only parking I knew. Both of them laughed as they knew that I could also park head-in. I added that it was the only parking I was familiar with in Iraq. Parking in reverse reduces the risk of hitting someone or hitting another vehicle. By reverse or combat parking, I could avoid backing out blindly into oncoming traffic or into the path of pedestrians. I was trying to explain and could go on and on, but they said that we should go because Lydia's son was waiting. I stopped my litany about combat parking, and we all dashed to the DFAC to meet the Air Force.

Meeting with the Air Force

The Three Musketeers joined the line of troops and contractors waiting in line to wash their hands. Once our hands were clean, we entered the DFAC and followed the chow line after showing our badges to the guard. Lydia always liked to have salad with oil and vinegar. She always ate healthily, and I wished that I could be the same way. I liked to have salad too, but I ate more carbs, and spaghetti was my favorite. Spaghetti had been my favorite source of energy since basic training and my source of energy to be a top placer in physical fitness training. Lydia and Imee were amused by my story, especially when I mentioned Drill SGT McCroskey and how he grinned like the Joker in Batman. I continued with my story until I remembered that this was a special occasion. So, I stopped talking, anxious to see the Air Force.

This lunch was special because we were meeting Lydia's son. We carried our trays to the table and started asking Lydia how she was supposed to meet her son. She said he would find us through the guiding spirit. We all said, "Amen," and finally, her son showed up. We were so happy to meet him and said what a fortune it was to have mother and son seeing each other on the battlefield, of all places. Her son agreed and introduced us to his fellow extraordinary Air Force officers. We all shared stories while eating and savored the American camaraderie in the battlefield DFAC. Most warriors eat fast, and Lydia's son was no different. Before we knew it, we had all finished eating and were saying our goodbyes to the Air Force and wishing we would see each other again. We all shook hands except for Lydia. She hugged her son, wished all of them safety, and told them to always be on guard.

Salsa Night

Salsa night at Camp Victory brought cheer to most of the workers at the zoo. The Three Musketeers liked terpsichorean salsa, so we thought we would try it for a couple of weeks. We enjoyed the trial dance and decided to stop as we heard more bombings outside the wire. The enemy was lurking in every corner, and we could not afford to be off guard. The heartless enemy was determined to make life miserable and was happy to distract from a salsa night. The Three Musketeers stayed out of Camp Victory for weeks and focused on work and cardio exercise at the SFG compound. Nowhere is safe on a battlefield, but at least we could get back to our HQ immediately if there were more bombings in the area.

Rumors

There had been rumors that American stores such as Ace Hardware, Walmart, and more would open at Camp Victory. I thought that was good news. The stores would help boost Baghdad's economy, and more nationals would have economic opportunities to sustain themselves. The coalition forces led by the United States provided their subsistence through humanitarian work and by training their Iraqi Army and the Iraqi Police. The plan to secure, stabilize, and rebuild Iraq would be costly, but dealing with the terrorists abroad would help to secure our homes in the United States. I agreed with this notion, but I also believed that there would be a substantial financial risk, as the violence that afflicted Iraqi society stemmed from an insurgency that had grown more complex and lethal since the occupation of Iraq had started in 2003 and had continued to escalate. This instability complicated meaningful political reconciliation among Iraq's religious and tribal groups, reduced the effectiveness of U.S. reconstruction and capacity-building efforts, and reduced the hopes and expectations of the Iraqi people because of a lack of adequate jobs; water; fuel; electricity; and most of all, peaceful living.

Our presence in Iraq should have supported the nation's tranquility. However, increasing the ranks of Iraqi forces and transferring security forces to them had not reduced violence. Instead, violent attacks had continued to spike, and our company continued to deploy linguists to help translate the noble purpose of our presence. Notwithstanding these efforts, the violence had led to significant combatant casualties. Additionally, there was concern about how the

United States would sustain the preventative maintenance of wear and tear on military equipment and the growing replacement costs. The collateral damage and strain on American forces alone had reduced the troop readiness levels and reserve personnel. It was stressful to rebuild and focus on helping the Iraqi government rebuild a sound economy with the capacity to deliver essential services, including linguists. Although Iraq's economy had grown and U.S. efforts had helped restore portions of Iraq's infrastructure, the poor security environment and mismanagement had diminished the overall results of U.S. investments. With all of the ensuing stress, was it better to deploy more American troops and civilian contractors to secure the Iraqis' future or to leave it to Iraqi people to secure their nation and sustain themselves without foreign intervention? My silent freedom weighed the two options. The United States to me had been like the movie stalwart superheroes from Universal Studios and Marvel Studios, saving humankind from the villains. The United States had always been generous in assisting the world through foreign assistance, such as giving aid to other countries to support global peace, security, and development efforts and providing humanitarian relief during challenging times or crises. It was a strategic, economic, and moral imperative for the United States and was vital to the country's national security. We also had our own battles that needed to be resolved, but that would not stop us from helping other countries. As it states in Romans 14:7, "We do not live for ourselves only, and we do not die for ourselves only."

THE SURGE

George W. Bush, 43rd president of the United States, gave a speech on January 10, 2007, that modified the strategy for helping the Iraqis carry out their campaign to put down sectarian violence and bring security to the people of Baghdad. The effort meant increasing American force levels and committing more than 20,000 additional American troops to Iraq. I was still serving on active duty in the U.S. Army when the President gave this speech addressed to the nation. I redeployed from Tikrit, Iraq, at the end of summer 2006. In May 2007, I filed my retirement papers after serving for 23 years in the U.S. Army. I wanted to serve further, but I had health issues, from which I recovered after my retirement. I was so passionate about serving the country again. I did not know how and where to connect in order to serve again, but God pointed me in the right direction, and I became part of the surge in Iraq.

I had heard that, in early 2008, three brigade-sized units would deploy to Afghanistan to support OEF. One of them was my one and only 101st CAB, the unit with which I had served and deployed in Tikrit, Iraq, "Wings of Destiny" from the 101st Airborne Division. I had also heard that our former S-1 officer with the CAB had been promoted to lieutenant colonel. I hoped that she would become a general in the very near future. With her combat experience and steadfast leadership, she was very likely to get to the top. She was the one I had called when our friend's team from Bastogne was ambushed and he had been one of the KIA. I searched for her contact number through my Army account and broke the sad news. She said that she suddenly felt tired after hearing the sad news. I said I would pray for him, and she said the same thing. Then, we both hung up, and I was so glad to have talked to her; I hoped that we would cross paths again. The phone chat memorialized our friend's passing. It also brought back memories of our journey and happy memories of sitting on a bed of oil in Tikrit, Iraq, and the flood rescue. The phone chat also reminded me of her caring attitude, sense of humor, and constant care for the soldiers. I was her non-commissioned officer in charge, and I took care of business in the office while she attended multiple meetings with the brigade commander.

The surge increased the number of both locally hired and U.S-hired linguists, as they deployed with the troops in the battlefield. Al Qaeda and foreign fighters loomed in Anbar province, the largest province in Iraq, straddling much of the west and sharing long borders with Saudi Arabia, Jordan, and Syria, as shown in the map. However, despite its size, Anbar province is sparsely populated. Iraq has tens of millions of people, but Anbar was home to only about 1.2 million people during the invasion led by the United States in 2003. Anbar has a number of metropolitan areas, and many Iraqis live in the relatively luxurious areas near the Euphrates River; however, much of Anbar is desert with a few roads running through it. A large majority of the inhabitants of Anbar are Sunni Muslims, and most belong to the Arabic-speaking Dulaim tribe. Anbar province was a Sunni stronghold that had long supported Saddam Hussein's efforts to remain in power. Anbar's provincial capital is Ramadi, which was under the U.S. military occupation during the Iraq war and fortified by U.S. Marines. Other important cities are Fallujah, the insurgents' den, and Haditha where the Haditha Dam and surrounding areas were initially secured by the U.S. troops as part of the invasion in April 2003.

Whereas the Sunnis live mainly in the western part of Iraq, most Kurds live in the eastern part of the country. I met mostly Kurds when I was deployed to Iraq as a soldier. Our interpreter was Kurdish, and the only reason I knew that was because he told me that he was a Kurd. The Kurds are our allies. I have deep gratitude and respect for Kurdish nationals because they served alongside our U.S. troops throughout the Iraq war and in other conflicts with the Islamic State. We have a strong camaraderie with the Kurds, and they are

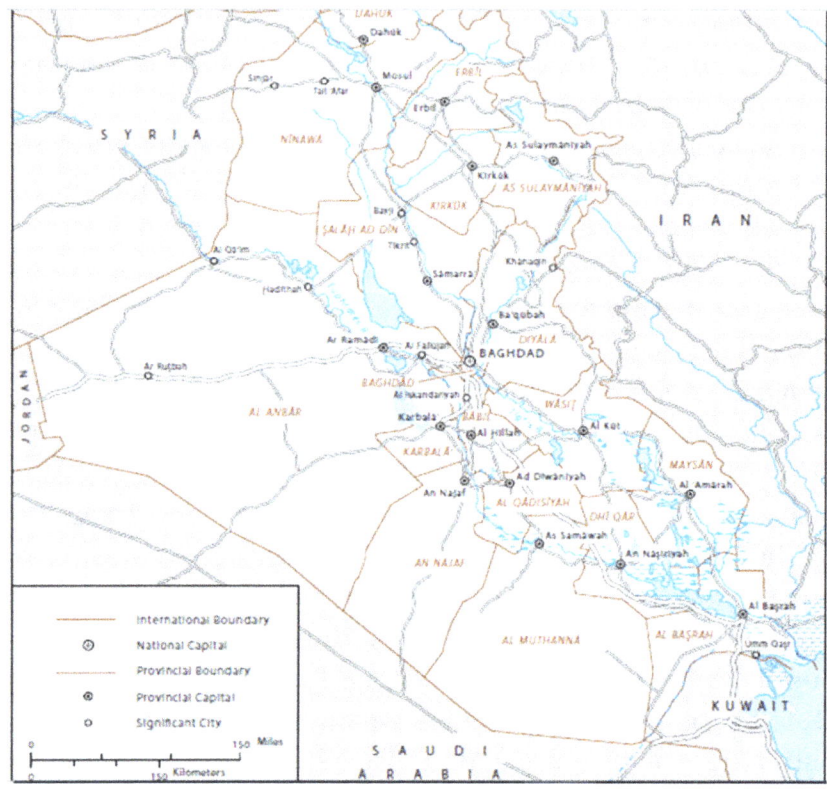

Al Qaeda lives in the province of Al Anbar.

dedicated to fight for the cause and even die for us. Most of us who served in Iraq will miss the Kurds, as they became like our brothers, just like our unit interpreter who fought with us. Recently, the Kurdish warfighters have been the main U.S. ally along the Syria-Turkey border; however, the Turkish government regards them as a terrorist group.

In his speech, President Bush stated that al Qaeda had helped make Anbar the most violent area of Iraq outside the capital. Indeed, it was, as al Qaeda carried out bombings and killings of innocent Iraqi citizens using chemical weapons, RPGs, and suicide attacks. They would also pay about $40 to anyone who would lay a roadside bomb, and even more than that if the bomb blew up and wounded or killed an

American. Al Qaeda had offered an ideological and religious appeal to strike back at the coalition forces. However, if their orders were not followed, they committed merciless torture and murder of the hired individuals and their families.

A captured al Qaeda document described the terrorists' plan to infiltrate and seize control of Anbar province. Such a plan would have brought al Qaeda much closer to its intention of taking down Iraq's democracy, building a radical Islamic empire, and launching new attacks on the United States at home and abroad. The surge in Anbar province convinced the local population to support the government and expel the enemy forces. It required coalition forces and Iraqi Special Forces to team up while conducting military operations and rebuilding and reconstruction efforts. The provincial government regrouped and reconnected with the central government in Baghdad, a critical step toward establishing a working relationship between the Iraqi capital and its provinces. Our linguists were a part of the peace process, as they helped translate English into Arabic and Arabic into English, in addition to interpreting documents pertinent to the mission.

However, we knew that the enemy was lurking in every corner of the area. I always had my helmet and Kevlar vest with me in my bed in case we were attacked by the enemy. I missed my M4, but at least I had some head protection against debris, as well as Kevlar, a material that can cushion, trap, and prevent a bullet from penetrating through the body in case of gunfire. The enemy saw us as Americans who deserved to be killed; they did not care if we were women or even mothers with young children at home. The American troops and American civilians were just another bounty for them.

My train of thought brought me back to my first deployment. I had seen people running rapidly around the fence at night and suicide bombers inside the military camps. We could not trust anyone; we did not know our enemies. We could not recognize them because the combatants were not wearing military uniforms. They did not have insignias; they could easily blend in with the civilian population. The enemy employed suicide bombers from among those we hired locally to undermine our determination to help the Iraqi people rebuild their lives and reconstruct their infrastructure. There had been several close calls on my first deployment to Iraq. Often, I saw local people staring at me from the other side of the fence. They could see anything inside the wire, which posed a danger to our lives. This was before the iron fences were replaced by tall blast walls to protect the troops, establish stability, and eliminate deadly threats, such as indirect fire from rockets, mortars, and IEDs. The walls were popularly known as T-walls. The iron fences were gone. The modern warrior was now secured by gray, featureless, and crudely built blast walls, especially in Baghdad. Camps were strewn with concrete barriers, walls, and guard towers. The T-walls reminded me about the rumor of establishing American stores in the Baghdad area. The T-walls would have been perfect to secure the American stores and other interests if they became a reality.

Change of Contract

As the company changed, we had to reach out to our linguists in the outlying units. The company sent us as its diplomats, initially in a large group to administer the contracts with our linguists and to offer some personnel services such as transportation if the site managers needed further assistance. Doug, other staff members, and I flew to Camp Taji, and the site manager picked us up at the airfield. After showing us our billets for the night, the site manager took us to the chow hall. There were short T-walls on the outside of the facility, a clearing barrel to clear weapons, and soldiers checking ID cards before the entrance to the chow hall. After showing our ID cards, I noticed that tons of MILVANs surrounded the chow hall instead of tall blast walls. It was the first time I had seen a DFAC not fortified by T-walls; instead, they used their MILVANs for protection. As we entered the facility, we joined the line of soldiers and civilian personnel waiting for available sinks at the hand wash stations that were set up for cleaning hands before entering the dining facility. In other camps, the sinks were set up outside the dining facility. It was a tradition in Iraq to wash hands prior to entering any dining facility.

The washing of hands before eating reminds me of a passage from the book of Matthew (15:1–8, 18–20) about the Pharisees and some of the scribes who had come from Jerusalem and gathered around Jesus. They saw some of His disciples eating with hands that were defiled, meaning unwashed.

> Now in holding the tradition of the elders, the Pharisees and all the Jews do not eat until they wash their hands ceremonially. So, the Pharisees and scribes questioned Jesus: "Why do your disciples not wash according to the tradition of the elders? Instead, they eat with defiled hands." Jesus answered them, "Isaiah prophesied correctly about you hypocrites, as it is written:
>
>> These people honor me with their lips, but their hearts are far from me.
>> They worship me in vain; they teach as doctrine the precepts of men.
>
> You have disregarded the commandment of God to keep the tradition of men.

Nothing that enters a man from the outside can defile him, because it does not enter his heart, but it goes into the stomach and then is eliminated." (Thus all foods are clean.) He continued: "What comes out of a man, that is what defiles him. For from within the hearts of men come evil thoughts, sexual immorality, theft, murder, adultery, greed, wickedness, deceit, debauchery, envy, slander, arrogance, and foolishness. All these evils come from within, and these are what defile a man" (Matthew 15:10–20).

I can think of many times that I justified a particular action, such as washing hands. According to my Bible study, in this passage, Jesus is essentially saying that we can do whatever we want regarding our consumption of material things. We can eat however much we want, drink however much we want, listen to whatever we enjoy, and essentially take in whatever we want as long as nothing evil comes out of us. As long as we do not gossip or curse, or no wrong deeds come from us, then whatever we take in is okay. Doug touched my elbow and asked if I was okay. I told him that I remembered what the Bible says regarding the washing of hands before eating and what Jesus said about people being defiled by what comes out of them, not what enters them from the outside. Doug said that I should pause what I was thinking and that we would talk about it later. In the meantime, we had to join the chow line before it got dark, and I yielded.

Camp Taji was set up under cover and before entering the chow line. The sign boards above the hand washing area showed unit signs and instructions for all to see. We then passed through another set of double doors to join the chow line. The menu for the day was written on the board, and the favorites were hamburgers, Salisbury steak, frankfurters, and chicken fried steak. Plates were provided for personnel wanting to bring food to their chus or workplaces. We decided to eat our food inside the dining facility, so we could observe the flow of traffic and appreciate Camp Taji's DFAC. I always felt alone when I did not see troops from the 101st Airborne Division. But I kept it to myself and focused on our mission to reach out to our linguists in the field. The site manager told us the story of Camp Taji. He said that the facility was also referred to as Camp Cooke and was one of the military bases located in the immediate proximity of Baghdad, being about 19 miles from Baghdad. Camp Taji was used by the coalition forces in Iraq, not just U.S. forces. It was located near the rural district of Taji, or Al Taji. The base was in the central part of Iraq.

The site manager laid out his agenda on how we could see all the linguists so they could sign their contracts. He said that there would be a town hall meeting in the evening and that this would be the best time to meet with them and then conduct our business. After dinner, we met with some of the linguists and completed the paperwork for their new contracts. They thanked us for traveling to Camp Taji and making it convenient for them to sign. We smiled, and we also thanked them for their courage in helping our troops make a big difference and improve Iraq. They agreed, and the rest of the meeting was for questions and answers. The site manager told us that we would see the others in the morning at another town hall meeting. It was just not possible to meet them all at one time because of their job commitments with their respective units. The linguists would then have the opportunity to sign their new contracts. Three days later, we had accomplished our mission and returned to our home base at the zoo.

When we returned to the zoo, we had a staff meeting about facilitating the process. The boss said that a decision was made to disperse us in small groups for the future administration of contracts. Not only was the idea clever, but it also leveraged the opportunity to reach out to all of the linguists assigned in the outlying areas and sign their new agreements before the deadline. Notwithstanding the fear of RPGs hitting our helicopter, my partner and I traveled to Al Kut, Samarra, Camp Echo, located in Al Diwaniyah, between Basra and Baghdad. Al Diwaniyah is flat as a tabletop, with green fields and groves of trees, mostly palms. I saw an expanse of green from the helicopter that was cut by several waterways, which must have been the Tigris and Euphrates rivers. I was told that the flood in Noah's time must have changed the course of the rivers, as they are seen today. I have been fascinated by the history of the Tigris and Euphrates rivers to this day, and I probably will be for a lifetime, because civilization began where the Tigris and Euphrates rivers meet.

We also went to Basra. I remember the time when our linguists showed me the border of Iraq and Iran, and it

The author before leaving Camp Victory, 2008.

was a chilling moment. Nevertheless, it was the journey of a lifetime that I will treasure for a long, long time, as I do not know when I might return to this place of conflict without our U.S. troops and our SFG. This deployment was an awakening and reinforcement of my faith, as well as my love for the United States and humankind. As my journey to other cities continued, it made me appreciate my freedom more as I learned Iraq's culture.

PART IV
THE FERTILE SOIL OF IRAQ

AERIAL VIEW

A view of southern Iraq from a UH-60 Black Hawk.

Al-Kut was our first destination in the southeast where I saw the border between Iraq and Iran. It lies along the Tigris River approximately 100 miles southeast of Baghdad. Al-Kut serves as a river port and agricultural center for nearby farms. My mission buddy said that explains the green fields that we saw earlier from the UH-60 Black Hawk. It was the first time I saw green fields in Iraq. I had been seeing distressed places, sand, and weathered dust storms since I deployed to this country. I had never seen such healthy vegetation, and even better, it lay along one of my favorite rivers, the Tigris River. The healthy vegetation vividly reminded me of the Garden of Eden, where everything was beautiful until the forbidden tree was violated. It was exhilarating to find vegetation in Iraq. For the first time, I could see the advent of peace in this country, and I told my mission buddy about my feelings. She nodded her head in approval.

Aerial view of the Tigris River.

My eyes feasted on southern Iraq's landscape. My karma was kicking in again as I saw the Tigris River streaming silently, and I felt that I wanted to come back here again under better circumstances and could not wait to see improvements. Iraq still suffered from violent and fatal attacks. Ten years after the U.S.-led invasion, suicide bombers in Iraq killed significantly more Iraqi civilians than coalition soldiers. Every day, we heard about bombing incidents hurting innocent men and mostly Iraqi women and children. Children were more likely to die than adults when injured by suicide bombs. The safety of the Iraqi people was one of the major blind spots of the Iraqi strategists and planners.

Later, after landing at the terminal, my mission buddy and I attended a town hall meeting with the linguists. We reiterated our purpose for going to Al-Kut, and the linguists

PART IV: The Fertile Soil of Iraq

said that it took courage to come to Al-Kut and that they appreciated our presence. They also added that they had a good site manager and that their units were taking care of them. They took us outside the town hall and showed us the border between Iraq and Iran. I promised that I would not cross over and that they were more important than Iran. They all laughed and said that I had a good sense of humor. Deep inside myself, I was very curious about what was out there, but I left it as it was and never looked back at the border.

The linguists also shared that Al-Kut was a regional center of the carpet trade. The Baghdad Nuclear Research Facility, which was looted following the invasion of Iraq in 2003, was located near Al-Kut. Al-Kut is best known as the site of a British defeat in the Iraqi theater of operations during World War I, which was fought from 1914 through 1918. In 1915, British forces occupied Al-Kut on their march to Baghdad. Military reversals led the British to retreat to Al-Kut, where they were surrounded by an Ottoman army. The British forces surrendered, and some 10,000 British and Indian soldiers were captured. I was fascinated by the story, and my karma had been active since we landed in Al-Kut. The Ottoman Empire was strong in the Middle Ages because it was big; controlled some very wealthy areas, such as Syria and Egypt; and had a powerful military that used advanced technology such as composite bows and gunpowder.

The site manager supplemented the linguists' story by telling us that, in 2003, the Marines adroitly made their attack against the enemy in Al-Kut and accomplished a strategic victory. They spent days in chemical suits and prepared for the possibility that the Iraqis had the capability and intention to use chemical weapons. The Iraqi conventional resistance in the south had seemingly melted away in the face of the Marines' advance. The Fedayeen and paramilitary forces had replaced the conventional defense as the main threat. The conventional defense had been vigorous but had proven ineffective. According to the site manager, the Iraqis had not made good use of their significant numbers of tanks and artillery systems and had not launched a much-anticipated bombardment with surface-to-surface missiles. Our linguists had been instrumental in maintaining successful communications and would continue to do so until units began to redeploy.

The site manager took us to the DFAC, where we continued our conversation about the war. We washed our hands and joined the chow line. I was dwarfed by the tall Georgian army soldiers as they picked up their trays and headed toward the food line. Our site managers had befriended some of them, and they joined us at our table. The Georgian army joined OIF as part of the U.S.-led coalition in August 2003. The initial deployment was a platoon of special forces and a medical team. Their presence in Iraq increased in 2004 and 2005 and peaked at more than 2,000 soldiers in mid-2008. Our site manager's presence made me feel more confident as we had a brief conversation with some of the brave Georgian troops.

On the Way to Camp Echo

We left Al-Kut the following day. From the Black Hawk, I took a picture of what seemed to be a huge unfinished mosque that was begun by the Saddam Hussein government in the middle of empty streets. In the late 1990s, Saddam Hussein invested hundreds of millions of dollars in three monumental projects to fortify his Islamic credentials and preserve his tyrannical legacy. Of these three projects, only one was completed: the Umm al-Ma'arik ("Mother of All Battles") Mosque, located in Baghdad, Iraq.

The unfinished Al-Rahman Mosque in Baghdad, begun by Saddam Hussein in 1998, was planned to be one of the largest mosques. However, the work was stopped because of the invasion in 2003. It remained unfinished in the Al Mansour district, a traditionally upper-class neighborhood in northwest Baghdad, located three miles from the Green Zone that was once home to diplomats and professionals who were wealthy enough to hire bodyguards. The Al-Rahman Mosque is surrounded by eight smaller, independent domes, which feature eight even smaller domes integrated into their walls. It is estimated to be 820 feet in diameter and occupies about 11 acres.

The unfinished Al-Rahman Mosque in Baghdad.

The third monumental project was the construction of the Grand Saddam Mosque, approximately two miles to the northeast, at the site of the old, abandoned Al Muthana municipal airport. The municipal airport was extensively damaged during the 1991 Gulf War. This aviation facility was used by the U.S. Army after the 2003 invasion of Iraq. It was planned to be bigger than the Al-Rahman Mosque and centered on an artificial lake, just like some of Saddam's palaces. It would have had a 200-foot-high dome, been surrounded by newly built educational facilities, and been about 980 feet in diameter. The construction of the Grand Saddam Mosque was halted at an earlier stage than the Al-Rahman Mosque, with only some central columns established.

The one and only finished monumental project, the Umm Al Ma'arik Mosque, is said to resemble the barrel of Kalashnikov rifles, manufactured in Russia, whereas the four inner minarets are shaped like Scud missiles sitting on launch pads. It is interesting to note that the Kalashnikov-like minarets stand 43 meters tall, which marks the 43 days of conflict with the United States that occurred during Operation Desert Storm. The inner minarets that look like Scud missiles are 37 meters tall and represent the year of Saddam's birth, 1937.

The more I saw Saddam's buildings that resulted from Operation Desert Storm, the more I was reminded of my former husband's sacrifice while serving during this war. In August 1990, Saddam Hussein ordered his army across the border into Kuwait, which was met with international condemnation. Kuwait was a major supplier of oil to the United States. Iraq's invasion was also an immediate threat to neighboring Saudi Arabia, a strategic ally of the United States and another major exporter of oil. If Saudi Arabia fell to Saddam Hussein's aggression, Iraq would control one-fifth of the world's oil supply. President George H. W. Bush, 41st President of the United States from 1989 to 1993, stated that this aggression would not stand. In the last months of 1990, the United States participated in the defense of Saudi

PART IV: The Fertile Soil of Iraq

Arabia in a deployment known as Operation Desert Shield. More than 500,000 American soldiers were deployed in Saudi Arabia in case of an Iraqi attack on the Saudis. When the troops were in place, I remember President H. W. Bush issuing an ultimatum to Saddam Hussein: leave Kuwait by January 15, 1991 or suffer the consequences of attack by a multinational force.

The date January 15, 1991, came and went with no response from Saddam Hussein. As a result, and true to his word, the next night, the President fulfilled his ultimatum, and the U.S.-led Desert Shield became Desert Storm. My former husband was one of the warrant officers deployed as part of the U.S. troops defending tiny Kuwait from Iraqi aggression. I requested to be deployed for this cause, but it was not meant to be. I stayed at home alone and became part of the rear command. I prayed for my former husband's safety and for the safety of all of the troops deployed in harm's way. It would have been easier to have deployed than to wait in anxiety and fear of the unknown. I remember going through sleepless nights and migraines while my former husband was deployed in support of Operation Desert Storm. One of the strategic and efficacious programs created by the rear command was the family support group, which brought together spouses of deployed troops. This alleviated worries as we helped each other, worked together, and volunteered our support such as lawn mowing, grocery shopping and delivery to spouses in need of such help, and many other philanthropic programs.

The spouses of Operation Desert Storm soldiers who were left in Fort Benning, GA, stayed abreast of the war and the status of their deployed husbands and wives. We frequently called to check in on each other and ask if anyone needed help. We constantly watched the news on TV and saw the bombings that battered Iraq's military targets for the next several weeks, destroying the same buildings that I would later visit during my last tour in Iraq in support of OIF IV. In response, Iraq launched Scud missiles at American military barracks in Saudi Arabia and Israel, of all countries, to trick the latter into joining Iraq's wicked cause. The Saudi Arabians were resolute and remained in opposition to Iraq. When the ground war began on February 24, 1991, my anxieties spiked, and my prayers became more intense. My church supported troops from the United States and its allies and prayed for their safety. I got more involved at church by being a lector, as well as in the Filipino-American Association and its activities to support the troops. Although the bombing lasted for weeks, the U.S. ground troops declared Kuwait liberated just 100 hours after the ground attack began. After the mission was accomplished, the troops gradually redeployed. We were going to pursue a withdrawal in 2003, 12 years later, after a series of four coordinated terrorist attacks by al-Qaeda against the United States on the morning of Tuesday, September 11, 2001.

Kuwait Liberation Medal

My former husband received three awards for his service in the 1990–1991 Gulf War: the Southwest Asia Service Medal for service in Southwest Asia from August 2, 1990 until a date determined by the Secretary of Defense; the Kuwait Liberation Medal, an award given by the Kingdom of Saudi Arabia for direct participation in Operation Desert Storm between January 17, 1991 and February 28, 1991; and the Kuwait Liberation Medal given by the government of Kuwait for participation in Desert Storm/Desert Shield from August 2, 1990 through August 31, 1993. Of the three, I liked the Kuwait Liberation Medal awarded by Saudi Arabia. It was also called the Saudi Arabian Liberation Medal and consisted of a silver star with 15 rounded points, with shorter rounded points between them, surmounted by a gold medallion containing a wreath tied at its base and a crown at its top.

Kuwait Liberation Medal, an award given by the Kingdom of Saudi Arabia for direct participation in Operation Desert Storm/Desert Shield between January 17, 1991 and February 28, 1991.

In the center of the gold medallion is a silver representation of Earth, over which is superimposed a gold representation of the Kingdom of Saudi Arabia. Above the gold medallion are the crossed swords and palm tree taken from the Royal Cypher. To me, it resembled the palm tree that the Three

Musketeers climbed at the zoo in Baghdad. When I think of Saudi Arabian countries, I think of palm trees. Beneath the gold medallion is a swallow and a tailed scroll with its ends folded back and pointed upward so they follow the contour of the gold medallion. The words "Liberation of Kuwait" are written in English with the same inscription above it in Arabic. There is no writing on the reverse side. Today, this is the only medal in its original box that survived our multiple moves as a result of permanent change of stations while serving in the active-duty army and multiple deployments in foreign countries.

Camp Echo

Camp Echo, or Forward Operating Base (FOB) Echo, was located near Diwaniyah, which is in the Multinational Division, Central-South area of operations. There were American soldiers on base. As the Sikorsky UH-60 Blackhawk touched down on the airfield, we dismounted and found the site manager waiting. After his welcome and greetings, the site manager gave us a ride to the site. I grounded my gear as soon as he showed us our billet and escorted us to the DFAC.

He stated that FOB Echo had been a Polish-run post since February 2005. There were about 2,500 troops from Poland, which was one of the countries among the "coalition of the willing," as it was called by former defense secretary Donald Rumsfeld. He paused in his monologue as we entered the hand-washing area and headed toward the available sinks. We dried our hands with the paper towels provided by the DFAC and entered the dining hall. As we joined the food line, I looked like a total stranger from a line dominated by troops from Poland. It was a glorious day to meet soldiers from the coalition of the willing as they joined us at our table. We had good conversations about the war and their willingness

PART IV: The Fertile Soil of Iraq

to support the effort. Our conversation was brief, as our site manager reminded us about our town hall meeting with the linguists. It was highly important that we see each one of them and sign their new contracts. We said, "do widzenia," which means goodbye in Polish. They responded with "trzymaj sie," which means take good care.

On day two at Camp Echo, my mission buddy and I thought we deserved a rich facial massage when we saw a shop in the shoppette, or small PX. We then indulged ourselves in a facial massage from one of the concessionaires and discovered that it was owned by a Filipina. What a coincidence! I always find people I know in strange places. The massage felt great, as I relaxed my body and felt the soothing massage for about 10 minutes. The massage came at a good time when we needed a break as we explored Camp Echo. My mission buddy and I laughed at the brief pampering and decided to continue relaxing at our billets. It was amazing to have facial massage on the battlefield. I took it as a good sign of economic progress, until the base was closed.

The next day, the site manager continued his story about the high-profile attacks including explosions involving the use of car bombs, suicide car bombs, and individuals wearing suicide vests that reached a peak in 2007. I agreed, as this was one of the reasons for the troop surge. However, because blast walls and hardened sites were constructed around the camps, and because kinetic operations were conducted against would-be perpetrators, the high-profile attacks slowed down and total vehicular attacks also decreased. Troop withdrawal might occur sooner or later as incidents continued to diminish. For now, though, our linguists continued to help the American units and made a huge difference in rebuilding Iraq, and the concessionaires continued to open until the drawdown. That night, I recalled the wonderful things I had seen from the air on our way to Camp Echo.

The author at Camp Echo Airfield in March 2009. Camp Echo is near Diwaniyah, which is located between Baghdad and Basra.

As I looked down from the air, I remembered seeing more well-irrigated areas around Diwaniyah from the nearby Euphrates River. It is oftentimes considered to be one of the most fertile parts of Iraq. I was in awe after seeing how fertile some parts of Iraq are, whereas some of our troops were serving in the middle of the desert and had to weather blinding sandstorms and go through the anxieties of Scud missile and rocket attacks.

The day came, and as soon as the last linguist had signed, the site manager took us to the airfield, and we were bound for our next destination. An Apache helicopter would take us to Basra, Iraq's second largest city, and the second to last leg of our tour to reach out to linguists in Camp Bucca. Doug had mentioned in an email that he knew the site manager in Basra, a former First Lieutenant in the U.S. Army. Later, the lieutenant would join another company and stay longer in Basra to fly drones, sometimes referred to as unmanned aerial vehicles (UAVs). I thought it was a very interesting and a smart change of profession.

The author at Camp Echo Airfield while waiting for her helicopter ride to the next journey.

The lieutenant was a very capable site manager, and it was not surprising that he would choose a more lucrative field. Evidently, he was a pilot and had a drone license to fly UAVs. Doug had mentioned that the lieutenant was knowledgeable and found his forte after OIF. He liked flying UAVs, and he liked to maintain a position as a ground-based controller. UAVs can operate with a range of autonomy: either under remote control by a human operator or under the control of an autonomous computer on board known as autopilot. Doug stated that drones originated primarily for military applications and that their use was swiftly finding many more applications in aerial photography, product deliveries, agriculture, policing and surveillance, infrastructure inspections, science, drone racing, and even smuggling. I was glad Doug was my friend. He was my walking Encyclopedia Britannica.

During our site visit, the lieutenant prepared an agenda and informed all the linguists to gather at 1900 hours to sign their new contracts. The linguists, in turn, asked their units to excuse them at that time so they could attend the meeting. It was a successful business meeting, and we obtained duly signed contracts that evening. We were supposed to have one more day before we completed our mission. The next morning, my mission buddy decided to find a laundromat and take care of her laundry. I decided not to go and talked to the linguists instead. Little did I know that the laundromat happened to be just right around the corner of the town hall, and we were surprised to find all of the facilities so close to the site. The linguists said in unison that they did not need a vehicle to go to the small PX, and we enjoyed that little joke. In the same conversation, they also told me that it would take only half a day to get to Iran from their location.

They hurriedly took me outside the building, and as we stood in a narrow aisle, they showed me the border of Iraq and Iran. It was a chilling moment to see Iran from Basra, and being in such close proximity to Iran was just unimaginable. I had goose bumps as they showed me the border, and I was curious about where the borderline was. However, I kept the question inside, and once again, I was grateful to my silent freedom. My drill sergeant's words echoed and reminded me that curiosity can sometimes kill you. I smiled at the thought and invited the linguists to go back inside the building and talk about something else. At that time, my mission buddy came back saying that she had successfully dropped off her laundry and would pick it up in the afternoon. I asked how much it cost her for same-day service. She responded that it cost almost nothing, and she laughed. She added that the

PART IV: The Fertile Soil of Iraq

urgent and compelling need would not prevent her from paying more if required. We both know that it was almost free, though, and we appreciated the benefits of the facility's availability.

I asked the site manager how far Mecca was from their site. He flatly said that it was 800 miles from Basra and that there were three ways to travel there: by bus, train, and automobile. Mecca is in Saudi Arabia, and Basra is in Iraq. Mecca is considered the city of Islam in Saudi Arabia, an oasis town situated in the Red Sea region of Hejaz, east of Jiddah in western Saudi Arabia. He quickly added that only Muslims were welcome in Mecca. I quickly responded that I was Christian, a Catholic by faith and, therefore, that I knew that I was not welcome to Mecca but that the question arose for people I knew. I was blessed to have known such people who respected my belief as I respected theirs. My mission buddy smiled and looked up at me as if saying, "Sometimes, you have to let go of your silent freedom and just say what you just said." I smiled back and gave her a thumbs up.

Samarra

The Blackhawk was ready, and my mission buddy and I were ready to go. We waved goodbye to the site manager at Camp Bucca and his assistants. It was such a productive visit, except for the site manager's smart remarks. Once again, we flew over scenic sites, and I pointed out the green vegetation to my mission buddy who also admired its beauty. As we moved along, we saw the damaged Al-Askari Mosque after an explosion in Samarra, which was situated 60 miles north of Baghdad. On February 22, 2006, members of al-Qaeda morphed into ISIS and infiltrated Samarra's Al-Askari Mosque. One of the Blackhawk passengers evidently followed the story and said that, after tying up sleeping guards inside, the fighters planted bombs that destroyed its famous golden dome. Such acts of religious vandalism started Iraq's slide into civil war. I contributed to the story and mentioned that I remembered taking a picture of the Al-Askari Mosque's golden dome during my trip to Samarra in 2005 with its exhilarating exterior from above. I found the picture and have shared it with my mission buddy later, and she was surprised and confirmed that it looked beautiful amid the conflict.

The Golden Dome.

Samarra is located on the east bank of the Tigris River in the Saladin Governorate. During the sectarian violence in Iraq, Samarra was in the Sunni "Triangle of Death." The Sunni Triangle is a densely populated region of Iraq to the north and west of Baghdad populated mostly by Sunni Muslim Arabs. The triangular area's points are near Baghdad, the southeast point; Ramadi on the southwest point; and the city of Tikrit, the north point, where my second deployment took place during OIF in 2004–2005. The area also contains the cities of Samarra and Fallujah.

As soon as we landed in Samarra, I asked the site manager, who had patiently waited at the terminal for our arrival, about the bombing of one of the holiest sites in Shia Islam. He replied that we must have seen the damaged mosque and added that no group had claimed responsibility for the attack, but they were concerned because it was near their site. The mosque was severely damaged, but fortunately, there were no casualties from the blast. However, the following day, more than 1,000 Iraqis were killed in the retaliatory violence. Because of the high-intensity conflict between Iraqi Sunni and Shiite partisans, it escalated to a civil war. It was fascinating to hear the site manager's story as he explained that Sunni and Shia Muslims used to live peacefully together for centuries. In fact, it had become common for members of the two sects to intermarry and pray at the same mosques. They share faith in the Quran and the Prophet Muhammad's sayings and perform similar prayers, although they have different rituals and different interpretations of Islamic law.

He continued further that the root cause of the Shia–Sunni conflict might have lain in victimhood over the killing of Husayn, the Prophet Muhammad's grandson, way back in the seventh century and long neglect of the Shia by the Sunni majority. Thus, Islam's dominant sect, which comprises roughly 85% of the world's approximately 1.6 billion Muslims, regard Shia Islam with suspicion, and extremist Sunnis have described Shias as nonconformists and apostates. His story ended as soon as we reached the site, and he told one of his assistants to show us our billets and ground our gear, after which he would drive us all down to the famous DFAC in Samarra. I was engrossed in listening to the manager's story and hoped it would continue after the meeting.

We did as we were told, and my mission buddy and I followed the smell of food. We washed our hands before entering the chow hall and joined the food line with our trays in hand. We rushed to look for an empty table as our growling stomachs anticipated a meal of grilled T-bone steaks with corn on the cob, rice, bread, and butter, followed by a drink of Coca-Cola. My mission buddy and I laughed as the site manager told us to slow down and to not choke. He joked that he might be a good historian, but he did not have experience performing first aid on someone who was choking. I stated that it looked like we had a happy group, and with that, we laughed together even more. We waved at the troops looking at us and who must have thought that we were crazy. We told them that Samarra was a nice place. They approved and waved back. Laughter is a great form of relief, and we all agreed.

We all left the chow hall with some corn chips and drinks in the pockets of our cargo pants for snacks later. I was surprised to see Aboona as soon as we reached the work site, and we hugged each other. Aboona is one of the nice linguists I met while training in Reston, Virginia, and again in Georgia. I introduced her to my mission buddy, and we chatted before proceeding to the meeting. The site manager introduced us to the group and explained our mission. They all nodded their heads, and we started passing out their individual contracts. We let them know that it was an open forum and that they could ask questions if they had any. It was a good meeting. Afterwards, my mission buddy and I attended to our personal hygiene in separate trailers and had a good sleep—at least I did, after the long flight and stressful jump trips we had taken over the past few days. I could not believe that I missed Baghdad already after just a few days. One more day and the mission would be complete, as soon as we landed safely at the zoo.

PART IV: The Fertile Soil of Iraq

Thinking about the zoo, Doug said that everything was about the same in the zoo and that he had been helping pick up mail from the hangar. That was good to hear, and I hoped that he was keeping my packages in a safe place. He was amused by what I said and assured me not to worry, as they were in the vault. He also said that he had planted a watermelon at the back of the fleet building and would see whether it would successfully produce watermelons. I could not wait to see it either and said that he would be honored by the Iraqis for such research. He sounded like he let out a guffaw at my email, and I followed it up with a hair-raising question about his rat. He said that Toupu and Sham from the fleet team had been feeding the rat with cereal and sliced apples. Mr. Arshad had also participated and brought some cheese every time he returned from the DFAC. I have reminded Doug to be sure to leave the rat with the zookeepers when it was time to leave Iraq. He said that he was never going to leave Iraq and that if he did, he would bring the rat home with him to Texas as his new pet. I told him he was insane, and he said he appreciated that. We shared a laugh through email, until I fell asleep over my laptop.

Back to Baghdad

The mission was complete in Samarra, and after I had said goodbye to Aboona, the site manager and his assistants drove my mission buddy and me to the airfield to catch another trip by helicopter. I told my mission buddy that we should have a grand pizza after this tour. She smiled and read my mind, extrapolating to a good plan for Luis's farewell party. We heard that Luis would be leaving Iraq soon to donate a kidney to his son. This was unfortunate, and we were going to miss Luis a lot. We talked about the plan, and with Delphine's blessing, we thought that maybe we could invite some of the linguists assigned to units near the zoo and some of his acquaintances in Baghdad to the party. We did not have time to nap in the helicopter, as we landed at Sather Air Force Base in no time. Baghdad is only 66 miles from Samarra. We found Sham and Mr. Arshad waiting in the vehicle as we appeared through the huge green tent. Mr. Arshad said, "Welcome back!" Remembering back to our first encounter, I replied, "Shukraan lika. Anah min aljayd ruyatuk maratan 'ukhraa" (that is, "Thank you. It is good to see you again."). Mr. Arshad let out a guffaw, and Sham seemed to be impressed by my Arabic. In his broken English, Mr. Arshad said, "Very good!" I replied, "Thanks, and I think it is not too bad." This time, Sham grinned and finally said, "Welcome back!"

After grounding our gear, we reported to Delphine and Luis and gave them an update. I heard that the other emissaries for the mission were also back, and Luis said, "Good job, everybody." I talked to Delphine one-on-one about Luis, and she sadly said that it was unfortunate for him to leave the zoo but heroic of him to donate one of his kidneys to his son. We were quiet for a few minutes, and after I had given her all of the signed contracts and a summary report in a Word document and an Excel spreadsheet, she thanked my mission buddy and me and said that we should rest a little and that we would discuss more about the plans tonight. I suggested pizza, and she asked, "What?" I then explained that we could give him a going-away pizza party by taking him out to North End Pizza at Camp Victory with all the staff members and having a plan to gather everyone in a catered party. Delphine approved and said that we would need to think of a theme for the party. I suggested, "What about a Hawaiian party?" Delphine asked again, "What?" I again suggested a Hawaiian party. I suggested we could decorate the room suitably for a Hawaiian party and pretend we were in Hawaii by wearing

Hawaiian outfits. Delphine said that it sounded like a good idea, and Imee started ordering Hawaiian ornaments from Oriental Trading Company for urgent delivery.

Dining Out at North End Pizza

It was pizza night with the HR crew. The excitement at North End Pizza, Camp Victory, expressed our anticipation of a successful gathering for what was to be the first and last outing with Luis. Apparently, Luis had a good rapport with the fleet team and requested a bus for our planned outing. Doug was able to produce a nice bus with Arabic lettering on the exterior. Our vehicle was not as opulent as a limousine in the United States, but it was loaded with the wonderful HR crew.

We occupied a whole picnic table with a limited number of seats outside the pizza joint. No reservation was required, and we were fortunate to get a vacant table considering that it was still daylight and on a weekend. There was typically a big crowd around North End Pizza on weekends, and coincidentally, the crowd was not as dense as usual today. We joined the end of the line waiting for pizza as soon as we disembarked and ordered several boxes of pizzas and sodas. There was still no alcohol on the battlefield at this time. We satiated our hunger with large slices of pizza. One pizza was half pepperoni and half Hawaiian with ham and pineapples. We had unbelievable access to pizza toppings even though we were in the battle zone.

The other pizza was for the vegetarians, such as Lydia and others. I looked around and noticed that the soldiers were looking for something to do on a Saturday night just like we were, and they were in line for either golden fries or pizza. Others might be at the MWR tent communicating with their families and friends or comfortably watching pirated movies

North End Pizza became a fellowship spot outside Oasis dining facility and a common attraction for visiting VIPs.

inside their chus. It was dark, and I did not see soldiers playing at Tumlin Field, where the sign read "'cause not all the fighting is done outside the wire." I really liked the wit of the sign and thought it was an appropriate way to describe the fights inside the wire.

North End Pizza was a blessing to the troops and was conveniently located near the DFAC. It offered the troops an outlet to fill their cravings with a different taste. North End Pizza became a fellowship spot and a common attraction for visiting VIPs. It would have been nice if they had served chicken smothered with different barbeque sauces. The owner said to watch the days coming ahead, as he was planning to have a variety of chicken wings and celery with dips. What big news! Just what the troops were looking for to spend their hard-earned money. Lydia laughed, but I told them about what had happened during OIF I, when we were observing what troops would do once the PX was established. We noticed that the troops spent money on bottled water, instead of drinking for free at the DFAC. Lydia sound amused and said that the troops have better taste, and we all laughed.

We ordered a variety of drinks with our pizzas such as Coca-Cola, Sprite, Dr. Pepper, orange soda, and water for the healthier people. The taste of Dr. Pepper fascinates me so that is what I ordered. It is a blend of 23 flavors. All flavors mixed together gives it a sort of pharmacy smell. I like the taste of anise and licorice in my soda, and Glen laughed when I said so. Imee, Lydia, and Delphine preferred water. I wish I could be as healthy as they are, but sodas like Dr. Pepper and Coca Cola serve as my medicine. I remember my mom telling me that soft drinks can get rid of stomach blockages. Jeff from the HR crew also added that Coca-Cola is a therapy for dissolving bezoars (or solid masses trapped in the gastrointestinal system, usually in the stomach). When it comes to stomach distress, many people have experienced that flat soda is a quick and popular remedy, usually in the form of cola, ginger ale, or clear sodas. Flat soda can help settle the stomach with its slight fizz and replenish fluids and glucose lost by vomiting and diarrhea. Arsenio from the HR staff laughed about what was said, and Luis said tactfully that maybe we should talk about something else other than bathrooms and panaceas. If we wanted to pursue the topic, then we would have plenty of time to discuss it back at the office. With that, everyone laughed and agreed to talk about something else that was more appetizing than bathroom sickness cures.

We could probably have utilized the dining facility for a night out. However, most of the staff members chose to have pizza at North End Pizza for the same reason as the troops. We wanted something different in a new environment. We chose North End Pizza over a relatively wide variety of restaurants such as Subway, Burger King, and Taco Bell. Besides, it was becoming monotonous going to the same place all the time. We used the DFAC every single day and would most likely continue to do so for the duration of OIF. Dining out was something we had not done before, so doing it as a pre-sendoff for Luis was extraordinary. I noticed that everyone seemed to be enjoying themselves on this special night, topped by exceptionally pleasant weather. It was not hot or cold. The August evening temperature was a gift from Heaven.

North End Pizza as it was getting dark and with almost no line.

It was also a full moon, so almost everything around us was still visible. Luis noticed the improvements in the site. He noted that it was amazing how the tall blast walls had miraculously saved our territory from the insurgents and helped us move more freely without fear of getting shot. The

PART IV: The Fertile Soil of Iraq

crew validated his statement and added that there was a new recreational bowling place where the troops could kill their boredom, as well as a chain of shops located close to the DFAC. Luis said that it was good but nothing compared with the building of shops in Camp Slayer that had an infinite supply of souvenir items from Iraq. The Three Musketeers were surprised and wondered how Luis knew about the hadji shops in Camp Slayer. We thought he never went anywhere else beyond DFAC. We thought our leadership did boring stuff and never explored the sites. One of the crew jokingly asked how Luis knew about the hadji shops, and Luis answered, "Remember, guys and gals, I was here even before you all got here." Everyone laughed and confirmed that it was true. No one could deny that Luis was indeed at the zoo before we all got there.

We asked where else Luis had been secretly traveling, and the Three Musketeers were hoping that he would not say that he had been to Baghdad International Airport (BIAP). Luis said that it would not be a secret anymore if he told us. Everyone burst into laughter once again. Luis was very entertaining, and while looking at him, I wondered if anyone else noticed that he resembled golfing champion Tiger Woods, although Luis had more hair than Tiger Woods. I told Lydia what I was thinking, and she agreed with me. We were wondering why Tiger Woods was losing his hair at such a young age. He should seek the advice of Luis about what to do with a thinning hair. With that, Lydia and I turned around to hide our laughter.

Everyone enjoyed Luis's entertainment. It looked like he had explored Camp Victory quite extensively and might have been to places that the Three Musketeers had yet to explore. Or he might just be bluffing us. He might have not been beyond the DFAC after all. The Three Musketeers smiled at the thought. Later, we agreed that Luis was a smart guy and that being retired from the military made him even more special. "Why does time go so fast when you are having fun?" asked Elaine. A smart guy replied that the excitement caused by the active pursuit of a goal was what caused us to perceive time passing so quickly. Because we were engaged in an activity that was focused on achieving a goal, time really did fly by as we were having fun. It got quiet, and all heads turned. The speaker was the designated bus driver, and Elaine said we better agree with him or else we could not go home. Then, there was more laughter. It was a well-spent evening with a special boss, and before we knew it, it was time to return to the zoo.

Remarkable Luau Site

The beautiful island of Hawaii is surrounded by crystal blue waters that only a tropical paradise could have, along with a landscape of volcanoes, emerald green rainforests, and tall sea cliffs. Such a description definitely does not apply to the zoo. As shown in the accompanying picture, our work site was a far cry from the beautiful island of Hawaii. First, the water was not a crystal blue sea, but a dingy canal. The climate was not tropical. Iraq has a hot, dry climate with long summers and short, cool winters. The climate is influenced by Iraq's location between the subtropical aridity of the Arabian Desert and the subtropical humidity of the Persian Gulf. There were neither volcanoes nor tall sea cliffs. The zoo did not have emerald green landscapes, but I appreciated the olive trees and bushes and the date palm trees, which I learned to like very much. Other than the palm trees, the zoo was far from Hawaii's perfect tropical zone.

However, I remembered Doug saying, "Nothing is impossible under the sun." I believed it and could not wait to receive the decorative items that we had ordered and that

Building at the zoo, Baghdad, Iraq. The water surrounding the location for the luau was not crystal blue seas but a dingy canal. However, the palm trees were real and perfect for the luau theme.

would make the conference room look like a real luau party. We were not in Hawaii per se, but we would decorate the place and make it look and feel like we were on the island of Hawaii. We ordered artificial hibiscus hairclips in different colors, leis for the ladies and garlands for the gentlemen, Hawaiian plates, tropical hats and fedoras with colorful bands, tiki-mask sunglasses, tropical table skirts, giant cardboard cutouts of a totem pole stand, tiki standee shindigs, and grass skirts. We asked everyone to wear Hawaiian attire such as print shirts or grass skirts over pants. The Three Musketeers thought that we would also have to order inflatable palm trees, although we had about 10 real palm trees: six trees in the back and the Three Musketeers' three date palm trees on the left side of the work site. Delphine blessed the plan, and although we did not want to see Luis leave, everyone was excited to go to the farewell luau. Delphine was in charge of the food, and the Three Musketeers were in charge of creating a Hawaiian atmosphere.

The word was out about the luau party for Luis. Most site managers confirmed receiving the invitation and confirmed that they would be attending, assigning their assistant site managers to take charge while they were gone. The company friends also responded and said they would come in Hawaiian attire. Based on the responses we have received, Luis apparently had many business and social acquaintances in Iraq. It was not surprising that Luis had lots of friends. He kept his promises, and he was the "master" in applying his military discipline. Everyone enjoyed working for Luis. We heard that the new Deputy Director would be coming from HQ, but the Three Musketeers were not worried as they were dedicated in doing their missions. However, Luis was like a father to us, and we would definitely miss him greatly.

The anxiety about receiving the Hawaiian decorations spiked as the farewell luau grew closer. The Three Musketeers scrutinized each package closely to ensure that the Hawaiian decorations would not get lost. This became the motivation for the next few weeks' mail runs. Luis, the celebrity of the luau, was as calm as he could be. There was no evidence of crippling anxiety. Doug mentioned that there was no need to lose sleep or to experience fatigue or any pain because he knew Luis was going back to his family and would help to see his son live. Doug added that the Three Musketeers seemed to be getting more anxious as the luau grew closer than anyone else at the zoo. He was right. But there were other members that were also getting excited, like Kathy and Robin from the Operations shop, as well as the stakeholders in the company. Iraq may have been the worst place to work, but what made it worth surviving were the great people that kept the company alive. Saying goodbye to Luis would be hard, yet we had to plan the perfect farewell for him. A big luau was a great way to send off a special person with great memories and to wish him the best. He was a family member who was leaving the family in Iraq. The farewell luau was a great opportunity to express goodwill and a spirit of human bonding with respect for Luis.

ARRIVAL OF THE HAWAIIAN DECORATIONS

Finally, the long-awaited Hawaiian decorations arrived. Lydia and Imee went for a mail run and found some boxes sitting in the hangar. They were excited to find big boxes and thanked all of the people who worked in the postal office. The boxes originated from Oriental Trading Company, which increased Lydia and Imee's excitement, and they returned to the zoo with exuberant news. Delphine and I were thrilled and called the men to help carry the boxes and place them in the conference room. Suddenly, the boxes brightened our hopes, as we had been waiting and wanted to start decorating as early as possible. With motivation, we could not wait to finish our work for the day and open the boxes. We were anxious to determine whether we had received all of decorations we ordered.

The Three Musketeers accounted for all the party kits and started separating the leis, garlands, and huge artificial palm leaves from the grass skirts; checked the loops and hooks in the grass skirts for easy attachment to the serving tables; straightened out the hibiscus clips and cute paper jazz toucan and the inflated Aloha sign; and accounted for the Hawaiian luau tiki torches, pineapple table centerpieces, paper lanterns, and more luau party kits. Lydia, Imee, and I stretched out four nine-foot paper garlands and hung them everywhere. We asked the tall guys to assist us, because we Three Musketeers were short people. From time to time, we went back to our desks to check on our work and ensure that everything was going well and that no one was requesting emergency assistance. The rest of the HR crew helped in pinning the cutouts to the walls and finding the best place to hang the banner welcoming everyone to Luis's luau.

Lydia sorts the first net for the luau as Imee rushes to help.

The author got the other nets for the luau.

We also draped the tall walls with nets and removable peel-and-stick palm tree wallpaper. Fortunately, they covered most of the wall, and Delphine was pleased with the creativity. We also ensured that there were extra plates, napkins, and utensils and that all of the tables were covered in colorful crepe paper. Imee played Hawaiian music, and it inspired the team to decorate the place and create an authentic tropical island atmosphere. Imee also played "Limbo Rock," and it totally rocked the house. We were happy that we got the tape in good condition and that it played perfectly. We could not wait to play it again during the luau and enjoy seeing people participate in the limbo, competing to see how low they could go. The limbo music was very motivating, and the folks demonstrated a sample of limbo. What a fun team! It would be hard for Luis to leave such a special, well-coordinated, and energized team, and knowing him, he would say that he wished he did not have to leave. A few hours later, Luis came by to visit the work in progress. He smiled with a satisfied look, as if approving that something had been done properly.

Farewell Luau

We helped Delphine in following up food orders and ensuring that they were ready for pickup. At a normal luau, the menu would consist of baked mahi mahi, banana bread, fun fruit skewers, banana or coconut cake, beef or chicken teriyaki, char siu, chicken adobo, chicken katsu, taro pudding, and perhaps a kalua pig. On the battlefield, though, such dishes were substituted by Subway sandwiches, pizzas, cheese balls, potato chips, fruit, potato chips, and dips. We had to appreciate what we did have, and it was better than nothing. Luis gave us an idea to let the guests know about the contents of the Subway sandwiches, so we printed labels, including labels for Italian, ham, chicken, meatballs, steak and cheese, tuna, and turkey Subway sandwiches, as well as sandwiches for the vegetarians.

Imee turned on the boombox and played "Isle of Aloha" music. It was very welcoming, and we saw smiling faces as people stepped up the aisle into the zoo's Aloha country. We thought about posting two ladies from HR at the front door to welcome guests, but then we thought Lydia and I should do it. With leis and garlands on our left arms, we followed lei etiquette. Each guest bowed his or her head, and we placed a flower lei or garland around the guest's neck, leis for ladies and garlands for men. Everyone accepted the leis and garlands and was proud to wear them throughout the luau. The hibiscus flowers with clips were given exclusively to the HR ladies, and they looked proud and pretty wearing them.

The first to arrive was Luis, the center of the party, and he showed up with a happy face. He was wearing a colorful Monstera Hawaiian print shirt with brown palm leaves, white hibiscus flowers, and small white and blue flowers with coconut buttons, a perfect shirt for a legend. He was always smooth-shaven, perhaps a discipline he had acquired in the military. He looked young in his shirt, and he received the first garland from Lydia along with a kiss on the cheek. The other guests arrived gradually, and most of them were dressed in Hawaiian attire, whereas others came in their regular workday attires and we adorned them with Hawaiian flowers. The guests felt relaxed as they sipped their near beers, the substitute for tropical drinks such as mai tais and the like, listening to Hawaiian tunes and watching the Three Musketeers as we danced the hula, which is the storytelling dance of the Hawaiian Islands. After our performance, everyone danced hula for a while until Luis announced that it was time for everyone to partake of the food and enjoy the atmosphere and have a good time.

After people had eaten, the big boss introduced the luau's celebrity and the reason we had gathered for the occasion. Luis stood up, and we all applauded as he centered himself on the platform. His speech was about the people on his team, others he dealt with every day, our military customers, and the company leadership. He said that he was sad to say goodbye and would miss some people, and then he paused as we all laughed at the word "some." In short, Luis was very articulate, and it was a good speech. The Company Director was the next speaker and echoed that we had a great, effective team that was irreplaceable. Luis would be hard to replace, but families are important, and Luis was leaving for a good reason. As I listened to the big boss, it reminded me that transitions are bittersweet, and I observed that people are resilient. I have seen people come and go in my 23 years in the Army, and people will be sad for a few hours, or a few days, and then bounce back and move on with life.

Luis was a tireless leader, always there to help and lead the way. He was rare, and I was glad to have had the opportunity of working with him on the battlefield and would do it again if given another chance. Because of Luis, I had accomplished my goals once again on the battlefield, in a civilian outfit. I was grateful to him for believing in me, my capabilities to work in his department, and the duties beyond the call of duty. I remembered him saying not to be afraid to make a decision and that fear was the biggest enemy of making decisions. To become more effective, get the fear out of the way. Overcome the fear of making wrong decisions. He told me like my father used to say, "Be the face of change. Without making decisions, you cannot expect personal and business progress. In a critical mission such as ours, make decisions based on your expectations. Be tactful, yet be fair." As he said those words, I felt like I was in Drill SGT McCroskey's world again, only at an advanced level.

After the speeches, Luis announced the limbo contest. Lydia and I looked for people to handle the bar and found some volunteers. Glen and Arsenio from HR volunteered, and before Imee played "Limbo Rock" on the boombox, Luis explained the rules: All contestants had to attempt to go under the bar with their backs facing toward the floor. When passing under the bar, contestants had to bend backwards. No part of their body was permitted to touch the bar, and no part other than their feet could touch the ground. They would be eliminated if their hands touched the ground. The contestants could not turn their heads or necks to the side. Whoever knocked the bar off or fell was eliminated. After everyone had passed under the bar in this manner, the bar would be lowered slightly, and the contest would continue. The contest would end when only one person could successfully pass under the bar.

Luis looked at Imee sitting next to the boombox, and Imee got the signal. To start the fun contest, Luis led the crowd by passing under the bar, and most guests followed. After the last person, Glen and Arsenio then lowered the bar, and Luis continued to pass under the bar, followed by Imee, Lydia, and me, and then the rest of the crowd. Everything was going well until the bar was moved lower and the "not-too-young" contestants were eliminated. Five contestants proved that they could go lower under the bar, and then it became four, and finally down to one, who became the Limbo Rock star and winner of a gift card from Subway. Everyone cheered and applauded. The winner happened to be the youngest person in HR. The Director bantered that the young winner had practiced before the luau event. Lydia quipped that youthful flexibility was over for us and had passed to the younger generation. Everyone laughed until Anita played the Macarena.

Suddenly, the bantering about the Limbo Rock star stopped, and everyone started dancing the Macarena. It seemed to be easier than limbo and was an awesome group dance. Delphine, Anita, the rest of the HR crew, and the Operations Team lined up and showed their dexterity in dancing the Macarena. It was great cardio exercise, and everyone seemed to relish the group dance. After the Macarena, most guests went back to the serving table for more food to re-energize. Next on the agenda was the presentation of farewell gifts to

PART IV: The Fertile Soil of Iraq   309

Luis. Each department gave their own souvenir memento, including the HR department, and was followed by the company leadership. Each group had a photo-op with Luis during the presentation, followed by more photo-op requests. Photos for men only, for ladies only, and then including all guests from the zoo. The party concluded by bidding goodbye to Luis. We all lined up to say "Aloha" and wish him the best. Maybe our paths would cross again. I said, "Amen."

A Bobcat and a Fox

A bobcat and a fox caught at the zoo in May 2009.

In May 2009, we heard a commotion outside the office, so we followed the sounds and saw a bobcat and a fox in a cage. Glen told us that men from work had caught these animals at the zoo and put them into the cage. I had never seen a bobcat before. What a pair! They are sometimes called wildcats. The bobcat in the cage was roughly twice as big as an average housecat. It had long legs, large paws, and tufted ears. It was brownish-red with a white underbelly and a short, black-tipped tail. I asked Glen if bobcats were dangerous. He said that bobcats do not attack people. Bobcat attacks are virtually unknown. Nevertheless, no one should ever attempt to touch or handle a wild bobcat or her kittens. Bobcats weigh between 15 and 40 pounds, which makes them small to medium carnivores.

I asked further what you should do if you encounter a bobcat. Glen was quick to advise that you should keep as much distance between you and the bobcat as possible and avoid running away because that could trigger a pursuit response. Also, if possible, spraying the animal with water would dismiss them, just like with regular cats. I said that that made a lot of sense. It was unbelievable having wild animals in the zoo these days when the zoo was occupied and with all the noise and gunfights outside the wire. Most innocent animals got caught in the war's crossfire, and uncontrolled looting and being eaten by starving citizens had killed nearly all of the animals from the zoo during the invasion. Luckily, the animals were rescued by the capable veterinarians of the zoo staff and compassionate U.S. soldiers, and a safe place was found for them in Baghdad.

I posed a follow-up question to Glen: Suppose I encountered a hyena on my way to the bathroom, which was approximately 150 feet away from my billet. Then what should I do? I had become dependent on his fatherly advice while we were talking about wild animals at the zoo. It was possible to have encounter with hyenas. Hyenas are not members of the dog or cat families; rather they are related to civets and mongooses. They are large doglike carnivores and have powerful builds with their shoulders higher than their hind legs. Glen said it was better to carry a stick or something to defend myself from the carnivores. It was possible to encounter animals after catching the two predators in the cage. It was a frightening thought, but I agreed with what he said. It sounded like it was time for the animals to reclaim their home now that the troops were gradually withdrawing from Iraq. Glen laughed and said it was highly possible.

On the contrary, I had seen some foxes in the springtime of the previous year and heard their howls whenever I walked around the perimeter at sunset. Foxes are small, doglike wild animals with pointed ears and nose and a thick tail. They are red and have dark front legs that look like small opera gloves. Foxes are known to be sneaky but look cute. We always heard the fox howls and barking calls at the zoo starting at 9:00 p.m. and they got worse during the full moon. We might be coexisting with some wild animals at the zoo, and we just did not see them during the day. Perhaps, the zookeepers would find them eventually and reunite them with the other animals.

Looking at the cage, I was surprised to see that the fox and the bobcat were not fighting at all. Instead, they looked tame and melancholy. After approximately an hour of commotion, it finally quieted down, and we all went back to work. The thought earlier about encountering wild animals on the way to the bathroom in the dark was disturbing. It reminded me of my crippling encounter with wild dogs at Camp Adder three years earlier, when I could have died from two dozen dog bites. The idea was not to run but to have some sticks or rocks or be ready to scream for help if possible. It was possible to see wild animals at the zoo. After all, we were occupying their homes during and after the invasion.

Heartbreaking News

The passing of a site manager in the Green Zone, commonly known as the International Zone (IZ) in Baghdad, Iraq, was heartbreaking news. John had retired from the U.S. Navy and joined the cause in Iraq in support of OIF. I kept hearing how smart and pleasant he was. He was very competent at managing our linguists in the IZ. The company leadership found him to be reliable and fair. The Green Zone was the governmental center of the Coalition Provisional Authority (CPA) during the occupation of Iraq after the 2003 U.S.-led invasion and remained the center of the international presence in the city. The Office of the U.S. Embassy & Consulates was located in the Green Zone. The Green Zone was completely surrounded by tall blast walls and barbed-wire fences with access available through only a handful entry control points, all controlled by coalition troops.

It was this high security that made the Green Zone the safest area of Baghdad, so that it was given the nickname of "the bubble." Like its neighboring cities, the Green Zone was frequently attacked by insurgents with mortars and rockets. I remember when the Green Zone was attacked with rocket and mortar fire almost every day from Easter 2008 until May 2008, causing a great number of civilian casualties. We were concerned about the safety of our linguists who were assigned at the IZ. We were glad that they were safe under the watch of our competent site manager, John. I remember the day when John tendered his resignation. We were sad that he was leaving us. Three days after he left Iraq, we heard that John had passed away from cardiac arrest, heartbreaking news to all of us indeed.

Paul's Visit and Tour of the IZ

One Sunday at the Catholic service at Camp Slayer, I thought I saw a familiar face, but I could not recall his name. It was embarrassing, but when he approached and said my name, it dawned on me that it was Paul, my classmate

PART IV: The Fertile Soil of Iraq

in the Advance Non-Commissioned Officer Course (ANCOC) at Fort Jackson, South Carolina, 14 years ago. What a small world! He had matured, and so had I. We hugged and both agreed that this was quite a place to meet away from home, in a battle zone. We chatted about what had brought us into the hostile nation of Iraq and also discussed the duration of our tours. We heard about the troop withdrawal from Iraq as decided by Congress. Paul suggested that we should see the countryside before leaving the country of conflict. The countryside meant a short tour to the IZ. I was excited to see someone other than troops from the 101st Airborne Division. Paul had retired from the U.S. Army and landed a federal job in southern Virginia. He said that he had tried his luck and passed the test and volunteered to go to Iraq to support their program and OIF. I told him he was lucky indeed, as he was assigned to Baghdad and not other cities in Iraq, like Mosul or somewhere in Anbar province, a hotbed for the bad guys. He said he was blessed and suggested that we should meet again. We then both agreed to meet the following Sunday after the Catholic service at Camp Slayer when I would be off duty.

Religious services were conducted inside the building surrounded by trees on Camp Slayer's lake.

Camp Slayer

Paul was smart. He found my base as I waited for him at the guard post. He was driving the company SUV, and I told him about my newfound skill of driving a manual transmission. He laughed and said that he worked for a logistics company, so it was not optional to learn to drive stick-shift vehicles. He mentioned that he was fortunate to have the skill and was easily hired for his current job. I asked Paul what set apart the experience of driving a manual transmission car. He replied that it was useful on the battlefield, where most vehicles have manual transmissions. Having the skill before deployment brought convenience and comfort. He said that the true essence of contentment came when you put your foot down on the clutch and changed gears. It was my turn to laugh, and I asked him about the biting point. Paul was fun to be with, and he reminded me of Bey, my teacher in driving a stick shift.

Paul paused a moment before answering my question, and then asked if I was referring to the point when a vehicle's clutch becomes engaged and the vehicle can start to move and change gears? I said that was exactly what it was, and he laughed. He asked again where I learned about the biting point from, and I told him about my training with Bey, my former co-worker. Paul laughed until we reached the gate guard at Camp Slayer. He said that my story was really funny and that he was glad that I finally had another skill up my sleeve, and we both laughed about it. We showed our badges to the gate guards from Uganda, and after looking at the pictures closely and looking back at us, they finally returned our badges and told us to proceed. We let out a big sigh of relief and moved on along the road to the IZ. Paul mentioned that he was leaving in four weeks, and I was speechless. I finally found words and said that God was good to have allowed us to find each other at Mass before he left the dangerous Iraq. He agreed and still could not believe we ran into each other in

Iraq. He asked who else from our ANCOC class was currently deployed. I told him I wish I knew. It would be great to get together and have souvenir pictures seeing Iraq in our rear. We both laughed as we stopped at the Victory Over America Palace, one of Saddam Hussein's showcases of vainglory.

VICTORY OVER AMERICA PALACE

Saddam's Victory Over America Palace in Baghdad was damaged by an American bomb during the years of conflict. It sits in the middle of Camp Slayer's serene lake.

The first building we saw was the monument to commemorate Saddam Hussein's dubious victory over the United States in the Gulf War in 1991. The front of the building evidently took a big hit, and the debris was still evident inside the palace where it was strewn all over the place. Paul said that the palace had never been completed in the first place because of economic sanctions and the bombing of the construction site. Its construction was brought to an immediate halt during the fearful invasion in 2003, when it received a direct blow. One of the construction cranes was left up to symbolize the ongoing construction before it got hit by the U.S. Hornets.

I was impressed by Paul's knowledge. He smiled, stating that he had toured the area previously with his work group. The palace complex was taken over by U.S. Marines after the invasion in 2003 and served as a military base instead of a palace for Saddam Hussein. It became a tourist spot for Marines as it was integrated into Camp Slayer. It is likely to remain incomplete forever, a lasting monument to a triumph that never was. The Victory Over America Palace now means Victory for America. Paul stated that U.S. soldiers could stand on top of the palace and fly an American flag to remember their loved ones at home. The troops could fly a flag in their honor and send a certificate home. When they got home, they then presented the flag.

RUINS FROM THE BOMB ATTACK

Ruins inside the Victory Over America Palace. This is some of the damage from an American bomb in the palace that Saddam had built to commemorate the first Gulf War, when his troops were forcefully expelled from Kuwait.

VAINGLORY

Paul pointed at the other palace sitting nearby, which is called the Victory Over Iran Palace, another symbol of Saddam Hussein's vainglory. The Victory Over America Palace was built behind and adjacent to the smaller Victory Over Iran Palace. Because of their location, the two damaged buildings are often mistaken as one. The ruined palaces sit on an undefiled man-made lake. The Perfume Palace and dozens of smaller vainglorious homes for Ba'ath Party dignitaries are also located at Camp Slayer.

Saddam Hussein's two architectural monuments to his hubris on Camp Slayer's lake.

Saddam's architectural hubris sits on the lake. I felt the effect of the water as Paul and I walked along the lake. It was more beneficial than living in the alluvial plain around Tikrit. Apparently, I felt more distressed when deployed in

Tikrit probably because it was hot and humid and had no lake and no greenery except for my green camo nets. I knew that there were beautiful sights in Tikrit; they just were not where we settled during the deployment. Seeing a lake in Iraq, particularly the Camp Slayer lake, combated the stress and put me closer to nature. Physiologically, my brain and body change when I am communing with nature, as I noticed when I lived on a house on a hilltop overlooking Ford Lake in Michigan prior to multiple deployments in the desert. My focus shifted from stress, and I felt more relaxed, which was good for my well-being.

PERFUME PALACE

Front door of the Perfume Palace on Camp Slayer's man-made lake.

Paul and I drove toward the circular domed Perfume Palace. The building was spared during the U.S. shock-and-awe bombing because of the building's circular dome, which closely resembled that of a mosque. It was Saddam Hussein's first wife's home. Perfume Palace was made functional by using it as the main HQ of the Iraq Survey Group (ISG), and it received minor damage from insurgent mortar fire during its occupancy by U.S. troops. The ISG's mission was to search and find weapons of mass destruction. When it left, intelligence operations started and continued through the closing of the palace by U.S. Forces-Iraq. Devices such as cameras were not allowed inside during the years when the facility was being used for intelligence operations. Before the invasion in 2003, the Perfume Palace was a brothel for Uday and Husay Hussein, Saddam's two sons, causing it to always smell of the perfume of their concubines. When the coalition forces captured the territory, the troops found numerous dead bodies on the palace grounds of people who were believed to have been murdered by Saddam's heartless sons.

The other side of the Perfume Palace.

More Ruins

After the Perfume Palace, Paul and I decided to drive to the next ruin, a damaged building that had visible debris from the bombing. Although the water prevented us from a close-up exploration, we managed to catch the cruelty of war. It must have been a wonderful building, just like the Twin Towers and the Pentagon before the multiple terrorist attacks. The attacks were horrible, and I can still vividly remember seeing people jump from the burning Twin Towers. These attacks caused emergency alerts and the deployment of military units to heal wounds. We were not going to be bullied, and we invaded Iraq in 2003.

The palace exterior and interior damage reminded tourists that an American bomb hit Saddam's Victory Over America Palace, just like the damage shown in these pictures. The front and sides of the building memorialized the Gulf War that commenced on January 17, 1991. My former husband had not said much about Desert Storm, but now, all of the ruins I saw in Baghdad spoke for themselves. Paul said that the American forces, under the command of Army General Norman Schwarzkopf, carried out a very effective strategy in defeating Saddam Hussein and his troops. General Schwarzkopf led the largest coalition of allied nations since World War II and a successful strategic operation by defeating the enemy, as demonstrated by these ruins. The U.S. Army, Air Force, Navy, and Marines struck at targets all across Iraq.

Tomahawk cruise missiles were fired from U.S. Navy surface vessels and submarines at large installations and communication centers. Paul said that the Iraqis gave it a fight too. It seemed that every Iraqi who could put his finger on a trigger had pressed down and would not let go. Most of the airbursts were below the U.S. Marine Corps F/A-18 Hornet strike-fighter aircrafts. The Iraqis fired surface-to-air

PART IV: The Fertile Soil of Iraq

More damage from an American bomb to the Victory Over America Palace, the palace Saddam built to mark the first Gulf War.

missiles blindly in the hope that one would hit something. Fortunately, the Iraqis did not do it right and paved the way for our troops to accurately fire on their targets, which became engulfed in flames and secondary explosions. All of the U.S. strike-fighter aircrafts returned safely after doing a great deal of damage to their targets.

Paul's statement about Iraqis firing blindly reminded me of my first deployment in Mosul, Iraq. Every single night, at about 8:00 p.m., the insurgents fired RPGs and surface-to-air missiles over our headquarters in the Rear HQ. Anxieties increased significantly as we took cover for protection. Every single night, we heard the swoosh of incoming rockets that continually arched over our roof, followed by nearby explosions. I remember putting both hands on my Kevlar and yelling "Take cover!" on a hard surface, as well as others.

I entrusted my life and soul to my Creator and prayed that He would protect my son if an RPG hit our building. I was hoping too that wearing the full battle uniform would secure us from bodily harm and that, if we got hit, the rescuers would not have to dig deep to find our bodies, dead or alive. I thanked God that we had strong hearts during the multiple attacks to have gone through such threatening missiles that haunt me to this day. I was jumpy to any sound, and it reminded me of the hostilities in Iraq.

During one of his visits to the Rear HQ, the commanding general (CG) stated that we pray that the Iraqis would never get their grids right and that we would continue to be safe on the battlefield. It was real, and we had to do our part and play safe. In my silent freedom, I expressed my desperation as to why the insurgents had not yet been traced or found. Every single night, the RPGs seemed to originate from the same point and swoosh over our HQ building and then land in the same location, every single time. Every night, we continued to suffer mentally and physically, and yet, none of the insurgents firing RPGs at us every single night were caught. Paul let out a big sigh and shook his head in disbelief. We both agreed that it was time to redeploy and give the Iraqis the opportunity to defend themselves, with our help.

I mentioned to Paul that my former husband was part of Desert Storm in 1991 and that all of these ruins reminded me of one of his heroic tours. Paul nodded his head and responded with admiration to the valor and dedication of the men and women who fought during Desert Storm. They had effectively expelled Saddam Hussein's troops from Kuwait, and the ruins that we saw that day were the successful manifestation of their resolute determination to strike strategic targets in Saddam Hussein's world. Air superiority, for example, was achieved on the first evening of Operation Desert Storm, striking strategic Iraqi leadership and communication targets

PART IV: The Fertile Soil of Iraq

and annihilating the elite Republican Guard in southern Iraq. It was also a good strategy to hit their air defense and artillery weapons as primary targets in Kuwait during the first weeks.

Paul was articulate and a good listener. I had known him for almost 14 years, starting in ANCOC at Fort Jackson, South Carolina. Besides, we were both U.S. Army Veterans, and so, we had a common ground to start with. I felt that I could trust him with my story. I told Paul that I had volunteered to deploy in support of Desert Storm but that my request did not go through. I wanted to join my former husband on the battlefield and make a huge difference for a great cause defending tiny Kuwait from Saddam Hussein's selfish desire and personal greed to dominate as the world's largest oil exporter. After hearing my statement, Paul looked at me with approval and said that I was brave and that it was quite a coincidence that we had a common story. He also admitted that he volunteered to deploy but that his request did not go through. He said that maybe it was meant to be, as he had young children at the time who needed their father's presence.

Paul continued with his newly acquired tourist knowledge and took me to the hill with radars and antennas. We climbed the hill, close enough to satisfy my curiosity. Paul kept reminding me that it was off limits and that we should leave as soon as possible, and I agreed. In my silent freedom, I wished that Imee and Lydia were with us. Perhaps we could have gone into the off-limits areas. I always appreciated my silent freedom for my rebellious thoughts that could potentially lead me into big trouble. During the week after my first encounter with Paul at the Camp Slayer Catholic service, I mentioned to Lydia and Imee that I saw one of my old classmates from ANCOC at Fort Jackson, South Carolina, and about my meeting with him the following Sunday. They were delighted that I had found someone I actually knew from home and wondered when they could meet him. I responded that I would have to invite Paul over. I regretted it now and wished we had brought the two ladies with us and explored Camp Slayer's architectural hubris and the areas that were off limits.

Mini Mosque

A unique view of Saddam Hussein's man-made Camp Slayer, Baghdad, Iraq.

My thoughts were interrupted by Paul's question and apology that we had to move on to the next place. We drove around Saddam Hussein's park and took a picture of the interesting architecture. Based on my recent trip, the structure resembled the top of the Mother of all Battles Mosque in northwestern Baghdad, the Umm al-Qura Mosque. Paul read my mind, as we were both looking for answers as to why a miniature mosque was built in the park. A mosque is a place of worship for Muslims. Any act of worship that follows the Islamic rules of prayer can be said to create a mosque, regardless of whether it takes place in a special building. Mosques are known to have segregated spaces for men and women, and this basic pattern of organization has assumed disparate forms depending on the region, period, and denomination. Paul and I agreed that help from a tourist guide would have been perfect in explaining the different structures built along the lake.

The mini mosque built in the park was unique indeed and captured both Paul's and my attention. I looked at the mini mosque's minaret and asked Paul what he thought of the crier. He looked at me and asked if I meant to ask about the prayers in the mic. I said with a smile, "yes," and we both laughed. He had not heard the crier since arriving in Camp Slayer, and I was surprised to admit that I had not heard it at the zoo either. I recalled hearing it in Mosul every morning. Apparently, prayers through the minaret vary according to location in Iraq. In Mosul, I noticed that it came at certain hours, about seven times a day.

I recalled the incident of hearing the crier for the first time. I was wondering where the crier was located until our interpreter explained that what I had heard were prayers coming from the mosque's minaret and that they came several times a day. He asked if I prayed, and I said yes, except that I prayed in private. He advised me to get used to the crier because it was part of the culture and, everywhere I went, I would definitely hear the prayers from different minarets. I thanked him for the information. He was one of the trusted Kurds embedded with my unit.

From our Mosul base, we could see residents right outside the gate. Now that our translator had explained to me about the criers, I shared the information with Paul. All mosques have minarets, and the criers call from atop the minarets in order to be heard by those around the mosque. Paul asked if the criers bothered me. I said no; as a matter of fact, I missed them when I redeployed the first time, and he laughed. It was a unique experience to know the culture of Iraq and to live the history. He looked at me the way I looked at Rosemary before with respect when she said those words in front of me and Bey. I continued to say that everyone is different and that one has to embrace those differences to have peace. Paul gave me a high five after I said it.

Flintstone Cave

I took a picture of what is known as Flintstone Cave on the lake. I could visualize the real Flintstone Cave from prehistoric times and wondered if it really looked this small. Paul and I were like small children and chased each other inside the cave. It had scrawled writings outside and inside,

The Flintstone Cave on Camp Slayer lake.

mostly consisting of the names of soldiers who had also toured the area. Astonishingly, Paul and I did not have to duck to fit inside. It was perfect for children and for short people just like us. Inside, the building was empty except for small rubble. There was no visible wreckage, and the small bits of rubble were a mystery.

JACK AND JILL

Paul and I remembered the song, "Jack and Jill," as we entered the next cave, named after said song:

Jack and Jill went up the hill
To fetch a pail of water.
Jack fell down and broke his crown,
And Jill came tumbling after.
Jack got up and home did trot
As fast as he could caper
To old Dame Dob, who patched his nob
With vinegar and brown paper.

The Jack and Jill Cave at Camp Slayer.

We entered the Jack and Jill Cave and remembered why Jack broke his crown, because of the narrow steps with no railings. This cave was situated next to the Flintstone Cave and was a perfectly matched co-location.

Paul stated that Saddam built the Flintstone and Jack and Jill Caves for his grandchildren, as an apology for killing their fathers. I cringed at this. He was about to tell me more about the swimming pool where Saddam Hussein hid the murdered bodies, but I asked that he tell me about it later to not ruin our wonderful expedition inside the man-made caves.

After a picturesque tour, Paul decided to take me out to lunch at Camp Liberty. He asked what I liked for lunch, and I suggested Subway sandwiches. He was amused by the idea, and we hurried like small children and arrived at a Subway restaurant, famished. I had a BLT sandwich with a Dr. Pepper, and he had a vegetarian Subway sandwich with a Pepsi. I forgot that he was a vegetarian. I recalled the times after class when Olivia, Paul, David, and I used to eat lunch together and Paul always ordered vegetables. We always cracked jokes and teased Paul that he did not look like he was vegetarian. He responded that I did not look like I ate meat either, and we all burst into laughter. Paul was very amusing as he told me that he could read what I was thinking. When I asked what he thought I was thinking, he confirmed that I recalled the day we had sat together with Olivia and David and teased him about being a vegetarian. It was amazing how he accurately read my mind. We both laughed, and he suggested that we go to the PX after lunch. I told him that as long as I was not going to drive, I would go with him. He laughed again and said that I was so much fun to be with. I told him he should tell that to Bey. He asked who Bey was but stopped as soon as we reached the PX, a different PX from where my friend Helen worked.

PART IV: The Fertile Soil of Iraq

THE PX

Paul and I noticed the soldiers having their pictures taken on a TIME magazine cover. They inserted their faces into a cutout, with "Person of the Year: The American Soldier" at the bottom of the cover. We were amazed, and I should have done it too while in uniform for souvenir. There were more troops as we moved forward, and I saw a familiar face. It was SGT Crump, who was one of my troops in S1 in the 101st Airborne Division (AASLT) with whom I deployed at Camp Speicher two years earlier. We hugged and chatted. She was a private first class (PFC) two years earlier and had been promoted since then. She mentioned that she was working for the CG as his secretary, and I felt so proud of her. She then invited me to visit so she could show me her office, and I assured her that I would. I introduced her to Paul, and SGT Crump said he seemed to be a gentleman. Paul appreciated the compliment and thanked her for it. We said goodbye and promised that we would visit her soon.

It was time for souvenir hunting. Paul looked baffled as he saw limited colors of t-shirts. We went through each displayed t-shirt, jewelry, mugs, hats, and many other items. I looked at a cartouche and the options of having my name engraved. Then we moved on to wooden boxes with Arabic lettering on the lid. We decided to take a break from window shopping and stopped at Cinnabon, the famous cinnamon roll place, and took the food to a covered area where everyone stopped to take a break.

The iconic Cinnabon has a blend of brown sugar and spices rolled up in the dough and is then smothered with cream cheese frosting. It comes in large, medium, and bite-size. The Cinnabon filled us up, and with its sugar content, it undeniably raised our energy to the next level, just what we needed to have more energy as we planned to be back at the PX and bazaars for more shopping sprees. The shaded area where we had just eaten the Cinnabons was the same area where two soldiers had been killed recently in an RPG attack. We were silent for a few minutes in respect to the soldiers who were killed there. The two troops were killed while they were taking a break when suddenly there was an RPG attack. There were several others taking a smoke break during that RPG attack, and the victims were the only soldiers who got hit. Talk about destiny. Paul said that he had heard about it and that it was unfortunate. I apologized for the sad reminder, and he said that it was no problem at all.

We went back to the PX, and Paul made a comment about the limited colors of their t-shirts. I asked if he was having decision paralysis on the colors, and he laughed. Evidently, there are not too many color options on the battlefield. You can choose either the light or the dark or the blue or the black. Then I spilled out my experience of two years ago when I was looking for a pair of sports pants and jackets. I ended up buying two sweatshirts in black and blue and two pairs of pants in the same colors. I resolved my color decision paralysis, and I got the best of both worlds. Paul laughed and repeated that I was so much fun to shop with. I said, "Amen!"

We finally decided to call it a day at 4:00 p.m. and Paul drove me home to the zoo. This time, he drove inside the camp and dropped me off at my building. I asked if he knew his way back out, and he said he did. We agreed to meet again the following week and follow up by email. At dinner time, Lydia and Imee picked me up, and we all went to Camp Striker. The guards teased me that it was my hundredth time to go out today, and we all laughed. Imee drove the vehicle, and we hurried to wash our hands and entered the DFAC. The food line was a little bit long, and we patiently waited in line. I had a hamburger and a Coca-Cola, while Lydia went for her favorite soup and salad. Imee decided to have hot dogs, chips, and water.

I was anticipating their questions, and they learned that all went well during my meeting with Paul. I told them that we went to Mass and had photo shoots of the different historic palaces in Iraq. My views of the Victory Over America and Victory Over Iran Palaces were irreplaceable. They were twofold memories: my former husband's participation in Desert Storm and my participation in support of OIF. I continued with my story that Paul and I went to the communication hilltop in the IZ but stopped at the boundary, where it was off limits. As expected, Imee ribbed me that I should have brought them along so we could have explored what was in the off-limits area. My eyes went big, and I said, "Not again!" We all burst into laughter. We remembered our dangerous escapades at BIAP, a yellow zone that was off limits and that could have resulted in big trouble.

Forbidden Intimacy

Paul had more news about the troop withdrawal. He was really keeping up with it, just like we were at the zoo. Before he talked about the troop withdrawal, Paul picked me up from the guard post where I stood waiting for the last 10 minutes. He arrived at exactly 7:30 a.m. I told him that he was a man of his word, and he gave me a gentleman's smile. We arrived in time for the Catholic service at Camp Slayer. The priest's homily was very inspiring. A homily is a sermon given by a priest after the scripture readings. Its purpose is to provide insight into the meaning of the scripture and relate it to the lives of the parishioners of the church. A homily may also be a long speech given by a deacon to teach a moral lesson. Paul and I both participated in the communion and waited until the priest gave his blessings to signify the conclusion of the Mass before following the crowd exiting the chapel. The host priest shook everyone's hand, and it was an excellent Mass, as always.

After Mass, I suggested to Paul that we have lunch at Camp Slayer's famous DFAC. However, because the dining hall didn't open for about three hours, he decided to take me to his chu. I entered his chu and admired how he kept it neat and clean. They had good living conditions. Everything was organized, and I noticed that he had one bed, meaning that he did not have a roommate. He had a mini refrigerator, a TV, and a chair, and I was wowed by his bathroom. It was surprising to me that a chu could actually have a bathroom attached to it, unlike my chu at Camp Speicher, in Tikrit, where the bathrooms were built outside the chuville. Evidently, the living conditions had not yet deteriorated before the troops started to withdraw from Iraq. I asked what patriotic deed he had done to deserve living in such a nice chu. He grinned at my joke, a grin that was far from Drill SGT McCroskey's grin, like the Joker in Batman. He humbly said that he just got lucky. I sat on the chair that was neatly positioned in the corner, and Paul sat on his bed. We stared silently into each other's eyes for about four minutes. Everything was silent, and it felt like we were the only breathing human beings at that moment in the world. It was an unbearable situation, and I was not about to surrender to a complicated relationship.

Paul was attractive and a gentleman. His silent stare was powerful and was generating inexplicable heat and virtually intoxicating my mind. Suppose he made an attempt to come closer and grab my arms? Suppose he made an attempt and told me how he felt and then tried to kiss me? I became breathless with the thought. Suppose he attempted to pull my top off and told me endearing words to convince me to submit myself and make love to him. Suppose he told me that he had been dreaming of everything like this since our first encounter at church and wanted to make love to me until midnight and would want to do it again? The intimate thoughts made me feel uncomfortable and hot under the

PART IV: The Fertile Soil of Iraq

collar. I felt hot like a fire alarm that was ready to go off. Suddenly, I felt I am going to have a fever. I believe that Paul noticed that my face was red, so he asked if I was alright.

I started to dig something from my bag. Something that I had been hiding since I saw him this morning when he picked me up. Earlier in the week, I decided to shop at Camp Striker for Paul's send-off gift. I remember him looking at t-shirts but being brokenhearted because there were so few colors to select from. He was such a simple man, so it was not too hard to buy a souvenir for him. I gave him a box with a manly red bow and told him it was a humble send-off present and a memory of the battlefield in Iraq. He was surprised by the gift and slowly took it from my right hand. With my silent freedom, it was a perfect distraction while we were sitting inside his chu staring silently at each other, just the two of us. I could not endure the heat any longer, so gift giving was a perfect distraction from such intimate thoughts. The private moment was too much to bear, and I told myself that there were too many challenges on the battlefield already and I did not need anything more to add to it. I was not up for a complicated relationship and did not intend to have one right now. Paul had good qualities, characteristics that would make him a good catch for a partner. I told myself that he was also a husband and a father, which decreased the calculation for a good catch. Once again, I thanked my silent freedom for thinking quietly.

I saw Paul's happy face as he opened the box and found three t-shirts of different colors and a mug that said "Iraq in my rearview mirror." He was like a kid and would probably be jumping up and down if I were not there. He unfurled the t-shirts. All three shirts had long sleeves, one in light and dark colors, one in a dark color, and the third in a light color. He was so happy and said that I did not have to do that. I was glad that we were talking about the gifts now.

The burning sensation was killing me, and I felt too weak to resist if anything actually happened. I told him that I was so glad that he liked the gift and asked if I could use his bathroom, to which he said yes.

I think he planned to show me his chu today as we waited for lunch. Everything in his bathroom was clean, and I smelled the fresh scent of the towels. The towels smelled like the Mary Kay cologne I sold in the United States as a consultant. It is a mixture of gin berry and icy bergamot with a shot of liquid oxygen accord, a mixture of fragrance such as vanilla, balsam, and labdanum or gum. It also has middle notes of a cool cardamom, with mint leaf and violet pepper, with bottom notes of wild birch bark and amber. It is indeed a masculine scent, and I like it. I put the scented towel next to my skin and briefly enjoyed its fresh scent. After a few minutes, I realized that I was in Paul's bathroom, and I smiled and shook my head. I thought it could be irresistible, but unfortunately, my brain commanded my heart. The excuse to use his bathroom worked perfectly like the distraction of giving his send-off gifts. I gave a big sigh and simultaneously flushed the toilet, so he would not hear me. I saw him wearing one of his new t-shirts with a happy face when I exited his bathroom. I asked him to try on the other ones to make sure they fit, which they did. Apparently, I missed something that he was hiding in his bathroom. He went inside, fetched a tiny red box, and stared at me. He told me to close my eyes first before he opened the box. He was not playing fair. I did not ask him to close his eyes when I gave him his presents.

He finally let me open my eyes, and the red box showed a simple necklace with a simple cartouche pendant. It was simply beautiful, and I asked, "Who is it for?" He said it was for me, in remembrance of him. I gasped and simply said thank you, and it was my turn to say he did not have to do that. He smiled and replied that he knew and then asked me to turn around so he could place it around my neck. I could not resist his touch on my neck. My heart was pounding, and my blood was racing as he gently placed the necklace around my neck. It was a struggle for me to bear his skin touching mine during his effort to secure the clasp together. I asked myself, "And now what?" I did not want to turn around and face him. I tried to be casual and pretended to reach for something from my purse as soon as he secured the clasp. Then, facing him, he said that it looked perfect on me, and I told him that he looked good in his new t-shirt. At that time, he said it was time for lunch, and I felt relieved, as my temperature gradually dropped. I needed to be outside the chu and to get some fresh air. We both got off scot-free from sin.

Suddenly, I felt more comfortable, and my breathing became normal. Paul laughed and said that I looked scared while I was in the chu. "You didn't think I would take advantage of you during those private moments, did you?" he asked. I poked his right arm and said that I was dauntless. He gave a guffaw and said I was really fun to tease and asked what I was hungry for at the DFAC. I said I was hungry for something that was energizing. It was my turn to defend myself, and it surprisingly made him stop laughing. It was my turn to laugh, and he blushed. He found a good parking space and did some fast combat parking. I wish I was fast like him at combat parking. He was really an expert at it.

I could feel him still staring at me as we walked toward the DFAC. I warned him to be careful with the rocks ahead. There were lots of rocks and puddles on the way to DFAC, and he should be looking where he was going. Luckily, we were walking next to each other, so I caught him when he stumbled over a rock and met his stare once again and felt his fast heartbeat. I told him, "I told you so. I should be the one you are rescuing in your arms." He teased that it could be the other way around. Nonchalantly, I let him go instantly and proudly said that it was good that I was quick and strong or he could have tripped flat on his face. He blushed and said he owed me lunch, and we both burst out laughing.

We showed our badges to the DFAC detailed troops as soon as we reached the guard post and then proceeded to wash our hands and used abundant paper towels placed next to the sink. We entered the dining hall, picked up trays, and joined the food line. Paul started talking about the troop withdrawal. It sounded interesting, and I welcomed the news. I would rather leave Iraq in one piece than be in the rubble of dead bodies, just like he was referring to earlier about the swimming pool filled with dead bodies. I told him that nothing lasts forever but the Creator, and he agreed and continued with his news. He led me to an empty table that was far from everyone else so we did not have to worry about eavesdroppers.

Troop Withdrawal

Paul spilled his news that Congress had finally approved a troop withdrawal and that we would witness redeployments of low-density and high-demand units such as the Special Forces Group. Some troops or contractors would remain in the country to train Iraqi forces as they sought to manage their own country. I told Paul that I had talked to Captain (CPT) Patrick Franklin from the U.S. Army one day about redeployment. He asked who CPT Patrick was, and I told him he was my youngest stepson currently deployed with Corps of Engineers in Balad, Iraq. He looked surprised and said it was great that my stepson and I were both in Iraq together. He asked if I had visited him and then caught himself when he realized that such a visit would be out of the question. It was strictly forbidden and impossible.

PART IV: The Fertile Soil of Iraq

Patrick had stated that they would be stopping at Camp Striker on their ground convoy to the south. I asked if we could meet just so I could get a glimpse of him. He was not familiar with Camp Striker, so he said he was not sure we could meet. There had been a lot of ground convoys, and it seemed unsafe to be rushing to beat the traffic. I was brokenhearted about not seeing Patrick at Camp Striker. I wished him safe travels and told him to watch the road and not run over IEDs. He said that he would take good care and have a security briefing before heading out toward Kuwait.

It had been six years since March 20, 2003, when OIF, the U.S.-led coalition military operation in Iraq, was launched. The immediate goal was to remove Saddam Hussein's regime and destroy its ability to use weapons of mass destruction or make them available to terrorists. As time went by, the focus of OIF shifted from regime removal to an open-ended mission of helping the Iraqi government improve security, establish a system of governance, and rebuild Iraq.

Paul continued that, as the war in Iraq wound down, security improved, and Iraqis started managing their own affairs, a new U.S.–Iraqi security agreement had gone into effect. On January 1, 2009, a U.S.–Iraqi bilateral relation had begun. The Secretary of Defense called it a watershed. It was a turning point indicating that American involvement in Iraq was winding down. The transition would include some 140,000 U.S. troops deployed in Iraq, in addition to civilian experts and U.S. contractors, who currently provided substantial support to their Iraqi counterparts in the areas of security, governance, and rebuilding. It was evident that lasting change in Iraq would take significantly more time. Although troops would be withdrawing, a limited number of personnel would remain to continue U.S. engagement with Iraq in the form of training its security forces and helping them defend their country, as well as fostering economic development for the Iraqi people.

Troop withdrawal had been a topic of debate in Congress since 2005–2006. We heard about it, and we actually thought we would be returning home for good in 2006. The 101st Airborne Division (AASLT) was deployed at COB Speicher in Tikrit during this time, and we were briefed on the plans for troop withdrawal. Instead, there was a surge of troops and contractors to support OIF II. Now, the time had finally come, and Paul was telling me to prepare for retrograde operations as all U.S. combat operations would cease by the end of August 2010 and all troops would be withdrawn by December 2011. I replied that the timeline sounded good, and I prayed for everyone's safety as they tried to go home.

Revisiting the Park

Paul finished his news report, and we left the dining facility to revisit the park at Camp Slayer. It was tranquil, and he mentioned that he came here often even before our encounter at church. It seemed to help his well-being and sanity while he was away from home. I said that it was a great thing to do and proudly told him that I was blessed with friends like Lydia and Imee and that we were far from bored. I also told him that one of our co-workers from Texas had found a fitting name for the three of us and called us the Three Musketeers. Paul laughed but did not say anything. I continued, telling him that the three of us all worked in HR and had the critical mission of supporting our linguists assigned to the outlying units. Each unit in Iraq has a linguist that helps in communication with local authorities, clearing hot spots, and other crucial missions as given by the higher command. Paul said that his unit had good linguists and that they were well taken care of in his unit.

PART IV: The Fertile Soil of Iraq

He then asked if I wanted another photo-op, and I nodded. We revisited the Flintstone and Jack and Jill Caves on the lake, and Paul posed for pictures. He was wearing a black fleece jacket, khaki cargo pants, and an arm band with his badge. He had grown his hair a little bit and kept his mustache trimmed just like he did at ANCOC. Paul looked like a perfect combat patriot dressed in civilian clothes. Later, he would be recommended for a Patriot Award at work because of his unselfish dedication to serving the troops. He was a true patriot and a hero. During ANCOC, he helped those who struggled with studying and with the PT runs. He spent late nights so that they, too, would pass the rigorous tests. With his diligence and leadership, all of the struggling ANCOC candidates passed and graduated, and they claimed Paul as their patriot and new hero. And here I was, with Paul one-on-one, in the strange country of Iraq exploring the beautiful secrets of Camp Slayer. Paul interrupted my thoughts and asked, "Are you going to click the camera?" I did not realize that I had kept him waiting as he posed next to the caves. We both burst into laughter and took more pictures by the lake.

We talked more nonsense business matters, and he asked when I would be leaving Iraq. "Hoping I will leave soon," I replied, although I liked this place and felt like I had been here before. "Really?" asked Paul sarcastically. Because I told the truth, he did not believe me. He said that I looked more like I belonged to the United States, originating from the Philippines, and we both laughed at his statement. Fortunately, he believes in karma like I do, and I did not have to explain much about reincarnation. Karma theories suggest that the realm, condition, and form of reincarnation depend on the quality and quantity of karma. For those who believe in rebirth, every living being's soul transmigrates or recycles after death, carrying the seeds of karmic impulses from the life just completed into another life and a lifetime of karmas. Therefore, Paul continued, the cycle continues indefinitely, with the exception of those who consciously break the cycle and reach the realm of gods, and those who do not wish to continue in the cycle. What a wake-up call. I needed to break the cycle and have peace. Paul continued that his grandparents used to say that no one knew if some of their pet dog were once human beings. I recalled my mother saying that to me, too, in one of her bedtime stories.

For the last time in the park, he looked at my neck and the necklace that he had given me earlier today under unbearably private circumstances inside his chu. He commented that it looked good on me but did not ask for me to please take care of it. I smiled without responding. I did not need to. We talked some more nonsense business matters and even talked about Olivia, our fellow ANCOC graduate. We had good fellowship and memories about the training and, fortunately, we all retired and landed government jobs. We both continued to communicate with Olivia. In fact, she was always the first one to send me a Christmas card every year. She worked in Arlington, and Paul worked in southern Virginia. David worked in Arlington as well, and I worked in Washington, D.C. Finally, Paul and I came back to our senses. We called it a day and hugged. I gave him a kiss on the cheek and thanked him for such a wonderful day. He blushed and drove me home to the zoo.

The Musketeers to My Rescue

I did not invite Paul inside my billet. I did not want to go through the same forbidden intimate thoughts I had in his chu and have a fever. It had been overpowering, and it got close. I got out of the car, said goodbye, and wished him a safe flight and to make sure not to leave anything behind, especially his new t-shirts and mug. He laughed and replied that he was not going leave anything behind but me. It did not sound right, but we both understood what he meant. Mutually we said, "See you on the other side." At that, he got out, and I was afraid of what he might want to do next. He said not to worry and that he was sad to leave this week. I replied that I knew, and that was why I gave him the t-shirts. He said he wished he could stay longer so we could talk some more business nonsense topics. I liked having a pure friendship with Paul, nothing else. I felt more comfortable talking to him and cracking jokes. I feared the consequences of his silent stare and thought that he was asking for a kiss.

My rescue from this forbidden intimacy came as Lydia and Imee said hello and interrupted Paul's show of affection. I introduced him to my friends, and we all started a good conversation. After all, we were all Army Veterans who met in a strange place like Iraq. Paul said he had nothing else to do for the remainder of the day, so he offered to drive us to the DFAC of our choice. We took him to the SFG DFAC and showed him the wonderful Special Forces hideout. He looked at it with awe, amazed at how lucky they were to be living in the palace. We all laughed and got out of the car to wash our hands outside the DFAC. Every DFAC is different, and the SFG decided to locate its sinks outside the wonderful building. Paul was surprised to see no one in the food line, and we got through it quickly. We sat near the TV, on which Fox News was talking about troop withdrawal. We decided to focus our attention on our food as Lydia and Imee became more acquainted with Paul.

Lydia was a live wire like Imee and asked how long Paul had been in Iraq. He said that he had been there with the DOD for just for 12 months and that he was leaving this coming week. Imee asked why he could not stay longer to join our group and work out together. My blood rushed through my veins as she made the suggestion, but I kept it to myself and pretended to focus on my chicken. The rush was even more so when Paul replied that he thought that he would do just that. Lydia said, "Really? That means we could visit the Al-Faw Palace and eat at Camp Slayer." I felt trapped by the ambush of words and continued to work on my chicken. I excused myself to get some dessert and a drink. I was thinking of a strategy to win the battle from the three friends I had left at the table while I helped myself to the fruit salad. The three of them were laughing as I headed back to the table. Lydia looked surprised to see that I was carrying three bowls of desserts and lots of sodas and asked if I was going to finish all of them myself. My excuse came in handy as I said that it was for all of us, but Imee said that they had already eaten theirs. I was trapped, and Paul helped me with the bowls and gently grabbed my arm and politely said to have a seat. I think he felt that my arm was warm, and he looked at me once again but did not say anything and continued to carry on with the topic at hand.

The next question made me blush, and the crimson was obvious to Paul, who continued to laugh at my friends' jokes. They asked how we met, and I told them about my encounter with Paul and the friendly meetings. Paul proudly talked about our ANCOC experience under the hot, humid weather of Fort Jackson, South Carolina. There were four of us in our group, and we diligently worked on homework. We studied and quizzed each other to prepare for the hard tests, which we all passed with flying colors. We always ate together and were inseparable. We enjoyed the training to the core, and the drill sergeant made us lead the troops and even sing cadence during PT runs. He added that the four of us had good cadences, and mine were original. With that, we all laughed, and I quietly teased, "With that, Paul, you are leaving tomorrow." But he was not paying attention as he carried on with the conversation. My silent freedom murmured that the

PART IV: The Fertile Soil of Iraq

near beers were not making him drunk, or were they? I think Paul heard me and said, "I am sober." Lydia and Imee said we both looked sweet. Paul and I laughed as we remembered how he teased me about it.

After lunch, Paul started to drive us home, but Lydia and Imee wanted to give him more of a tour of the compound, and Imee's witty conversation kept us all laughing. They showed Paul the olive trees, the gym, and other facilities. I remembered Bey when we passed the olive trees, and I smiled as I remembered our juvenility as we almost got caught stealing some olives. Paul was fun, and I agreed that he should stay longer so we could all leave Iraq together and then view Iraq in our rearview mirrors. He said sarcastically, "Really? Do you really want me to extend?" Suddenly, I regretted what I had said. I did not really want him to extend. Lydia and Imee said they were off today, too, so we should go to the PX at Camp Striker. Paul was accommodating and acquiesced to Lydia and Imee's whimsical ideas. They liked him and wanted him to stay longer.

My friend Helen from Ethiopia worked at Camp Striker's PX, and we chatted on our way in. We looked around and left as fast as we had come in. We went to the bazaars and the Green Beans Coffee shop. Imee said that she was going to buy us all coffee, but Paul said no, it was his treat, and we all yielded. If Lydia and Imee could do anything to keep him in Iraq, they would. There was a lot to do at Camp Striker, and Paul seemed to enjoy our company and was determined to overindulge Lydia and Imee on about anything they asked for. He had spoiled them for about four hours already. Imee showed him the DFAC and the gym where we always exercised in the morning, along with the shower building. Paul looked at me at the word "shower." I gave him a "not a chance" look, and he laughed. Could he really extend his tour? How easy would that be? I was thinking of the feasibility of extending a tour, especially when troops had a timeline to withdraw. I convinced myself that he was not going to extend, and the thought made me breathe easier.

It turned out I was right. Lydia and Imee suggested that we should see him off at the airfield. I smiled at the suggestion, and my silent freedom said that, yes, I could do that. Everyone agreed, and we jotted down Paul's itinerary before he drove us home to the zoo. It was a productive day, and we enjoyed Paul's doting company. He did well for my friends that day, and he gained their trust during our short time dining and shopping together. He was not selfish, and I liked that quality in him. He now had more friends in Baghdad but, unfortunately, he had to leave soon, so he would not get to enjoy our pleasant company and explore the city with us further. On the other hand, I wanted to see him depart, so I agreed to come to the airfield. He looked at me intimately when I agreed with Lydia and Imee that we would see him off at the terminal and that we would see him on the other side. We reached the zoo, and the guards let us through. Paul had become a familiar face to the guards in such a short time. We all got out of the car as soon as we reached my billet and chatted. He said he was happy with the gesture and expressed his gratitude for our showing him the SFG compound and having lunch with him. Then Imee and Lydia said that it had been such a pleasure to meet him and a shame that he would be leaving in a week. Paul was speechless for a minute and said he wished that he could stay longer. He then got in his car, waved goodbye, and said that we would see him again at the airfield terminal in a few days.

As expected, Lydia and Imee said Paul was such an attractive gentleman. He was such a pleasant treasure and a grand protector. I asked what that implied. They both laughed and said that it meant that he was good and a champion. Their statement echoed SGT Crump's statement earlier at the shoppette that Paul seemed to be a gentleman. Lydia posed a stunning question: "Did he kiss you? Did something happen in the chu?" I gave a big sigh and replied with an emphatic "NO." We got out of his chu before anything happened. But yes, I suffered silently from physical and mental thoughts, and we got very close to intimacy. Lydia and Imee asked, "Why not?" With that, I gave them a look and said, "I think we will leave it at that, and we will see each other tomorrow morning at the office." They both laughed and teased about hoping that I could get some sleep to suppress my feelings. I resolutely replied, "I will!" With that, we parted ways at the zoo. Lydia and Imee went to their respective billets, which were about 50 feet from mine.

SFG Complex

I gave this photo to Paul because he liked the SFG complex. It was taken from the hill, and the large building beyond the arch is the infamous Al-Faw Palace resort complex, HQ to the U.S.-led coalition forces.

The Kiss

I emailed a picture of the SFG complex to Paul before he left Iraq, and he said that he liked and appreciated it. We communicated more by email, and he said that he was all packed and ready to go. On the day he was scheduled to depart Baghdad, he had his backpack and was wearing one of his new t-shirts. I was wearing my exquisite necklace, and he smiled upon noticing it. Before he proceeded to the manifest station inside the huge green tent and left Iraq for good, I hugged and kissed him lightly on his mouth. I thought it was the least I could do to make him feel good as he departed Iraq. This surprised him, and Lydia and Imee were quick to take a picture. It had been a wonderful encounter, and we wished him the best and said that we would see him on the other side. I was stunned when Paul set his backpack on the ground and gently grabbed and hugged me as close as he could and gave me a long kiss. At that moment our hearts were synchronized, but I could not tell whose heart was beating faster, his or mine.

I felt trapped. Suddenly, I noticed that my arms were around his neck, and then, one of his hands cupped my chin as he gave me another passionate kiss. I almost melted in his arms, and I had to catch my breath when he finally let me go. It was unexpected, and he did not mind that Imee and Lydia were watching. I believe that they turned around to cover us from the other people. He winked and whispered something that I did not expect to hear. I was still in shock, and all I understood from what he said was to check my email.

He picked up his backpack and told Imee and Lydia that they can turn around now and gave them friendly hugs too, but not as close as the hugs he gave me. He looked at me one more time, touched the exquisite cartouche on my neck, and with my chin cupped in his hand, he gave me another tender, passionate kiss. Imee and Lydia covered their eyes. It felt like it would be a long night and Paul might not make it to his flight, which, to me, would be a bad idea. It was challenging to say no, when I meant to say yes. But sometimes, it is what it is. I had to suck it up. Softly, I told Paul that I would be alright and that I would always remember him, but that he should leave now so he did not miss his flight. He gave me another silent stare, and with the same hand, he gave me a final, tender kiss. He said, "Take good care," waved goodbye, and looking at the three of us, uttered the familiar words, "See you on the other side." I could not remember how many times we had uttered those words since last weekend, but his hugs and kisses shocked me like a thunderbolt and made me feel weak. Paul looked happy as he entered the green tent before boarding the aircraft. Lydia said my face was as crimson as a rose and asked if I would be okay. Everything was like a dream, so Imee pinched my hand and said everything was real. We stood outside the tent for about 15 more minutes until we saw Paul come out on the other side of the tent and head toward the aircraft. Imee said something about my facial expression after the kisses. When I asked her to explain, she said I looked funny. We all burst into laughter as Imee said she could not wait until she went on vacation in a few days and was able to see her husband. We shared some girl talk as Imee drove us to the Sather Air Base dining facility.

    I had not kissed anyone for a long time, and Paul's affection undeniably made me feel good. What Imee said about the "happy hormones" was true: Being in love is somewhat scientific. She explained that oxytocin and dopamine lead to feelings of affection and euphoria, and kissing releases serotonin, which causes a surge of positive emotion. Love also lowers cortisol levels, which explained why I felt more relaxed and good all around. I asked Imee how that applied if there was no chemical connection or if the feeling was not mutual. She replied that, in that situation, there would be a disconnect, as the happy hormones do not work in isolation, but in concert. You would become detached and would feel uncertainty. The hormones would just run away. We all burst into laughter after listening to Imee's humor, and we called her "the Master," as she expressed her expertise in romance. She repeatedly said that she could not wait to go on vacation and be lovey-dovey again with her husband. We all laughed again. It was productive girl talk as we headed home to the zoo with full bellies.

Photo-ops with Lydia in Al-Faw Palace

Paul's surprising show of affection at the terminal made me a wimp, and I had many sleepless nights that week thinking about it. I showed up to work sleepy. I finally succumbed and talked to Lydia about it. She giggled at my juvenile insanity and suggested that we get some fresh air at the Al-Faw Palace resort complex and take some photos. It would also cure our yearning for Imee after she had left for R&R in the United States, she said. I welcomed the idea wholeheartedly, and we attended the Catholic service at Camp Slayer before touring the palace and then having lunch at the Sports Oasis DFAC later.

We met a couple of Australian troops and asked if they would like to take pictures with us in front of the palace. They looked delighted at the idea, as they said that they were also acting like tourists before redeploying to Australia. The photographer said, "Say cheese," and we all complied. We chatted with the Australian troops after the photo-op before finally bidding each other adieu.

PART IV: The Fertile Soil of Iraq 349

The author and Lydia with soldiers from the Australian Army, who agreed to have photo-ops with us.

After the Australian troops had gone, Lydia and I noticed two identical bunkers in front of the palace, which we decided would be a good backdrop for our next photos. It takes a team effort to fortify a secured bunker, and based on our inspection, they were solid and seemed to provide good protection. We encountered U.S. troops by the bunkers and took pictures with them as well. Everyone seemed happy to know that we were finally leaving Iraq and would have some pictures to remember what we had seen.

THE BUNKERS

The bunkers brought vivid and horrible memories of my previous deployments when I was still wearing my military uniform. Back then, we jumped into bunkers every time the insurgents remorselessly attacked with RPGs. Our foxholes were chest deep, about 6.5 feet by 6 feet. The standard size is 6.5 feet by 3 feet. There were several soldiers in each of our foxholes. Our tactical bunkers were in the form of dug holes that were covered by plywood, thick layers of dirt, and sandbags and were not as clean or fancy as the two bunkers on the palace compound.

The author and Lydia outside one of the two clean, identical bunkers in front of Al-Faw Palace.

I closed my eyes as I recalled the hard days of war. During the first OIF, we looked for strategic locations to dig several foxholes. We jumped into those foxholes several times and stayed in the holes for about 30 minutes until we heard the words, "All clear." I had seen RPGs exploding outside the perimeter. Each RPG provided a vigorous outward release of energy that would have been unimaginable if one had hit our living areas. Only God knows if we would have survived under those circumstances. The scar of war is indelible. Although the war might be over, the memories remain forever. Who would forget the troops we lost from the ferocity of war? What would cure the disorders from ambushes and gunfights? What would erase the anxieties and excruciating fear of what would hit next? I thank the Lord for the cleverness and ingenuity of our leadership and troops for being creative and securing the modern troops with living areas fortified with tall blast walls and sandbags. I also thank the Lord for the blessings of fortified vehicles securing our infantry as they patrolled borders and for saving Patrick's life from the blast of an IED.

As the war in Iraq appeared to be winding down and as the Iraqis increasingly sought management of their own affairs, it was proper to build fancy bunkers like those at the palace. I remember Paul's words as he reported that President Obama had stated how the war in Iraq would end and announced that, by August 2010, some of the military mission would transition to training and advising Iraq security forces, conducting counterterrorism operations, and providing force protection for U.S. personnel deployed in Iraq. It reminds me of some of our courageous site managers who decided to stay in Iraq to this day, training and advising the Iraqis on how to defend themselves.

The bridge from the road to Al-Faw Palace was about one-quarter of a mile long, with an island between two pathways. Only authorized vehicles could park near the palace. Everyone else parked in a dirt parking lot and walked across the bridge. It was Sunday, and lots of troops from the coalition were spending their downtime at Al-Faw Palace. We all thought alike: Take as many memories as you could prior to the troop withdrawal. After all the troops left Iraq, who knew what would happen to the palace. Would the future generation remember that it had been the U.S.-led Command Center during OIF? The coalition forces preserved it well and secured it from the insurgents. Were the U.S. forces going to be part of their history? I knew that we would be remembered in U.S. history, but would freeing the Iraqis from Saddam Hussein's tyranny be remembered in Iraq's history? Whatever transformation project the Iraqis had in mind after the transition of power from the U.S.-led coalition to the Iraqis would be up to them. If the existing palace and surrounding mansions inside the complex, embellished with beautiful chandeliers, marble, and exquisite bannisters, were used as colleges and facilities, they might remember that the

The author and Lydia in front of the Al-Faw Palace.

Big Eagle landed in an effort to rescue them and help them gain freedom from Saddam Hussein's tyranny. To me, freedom matters, and it is a gift from God to have a government that helps promote inalienable rights, the pursuit of happiness, freedom from want, freedom to make our own choices, and freedom of speech and religion.

AL-FAW PALACE'S JADE GRAND ENTRANCE DOOR

Entering the palace was a delightful experience. Lydia had two reasons for visiting the palace: to see SGT Crump and to tour the infamous palace. So we hurried until we noticed a tall front door that seemed to be an impenetrable entrance made of jade. We decided to take a photo. I had never seen a door like it before. Lydia also seemed to be impressed, and we decided to stand next to a palm tree in front of the building and spent a few minutes admiring the architecture. Passersby were similarly curious and decided to take photos with the jade door as a backdrop. I told Lydia that I was as tall as the first bar on the door's left side and that she was as tall as the bar on the right side. We both laughed and waited for a few more minutes before proceeding through the door.

As soon as we entered the magnificent palace, a plaque on the wall caught my eye. It described the Al-Faw Palace and its surroundings. Specifically, the Al-Faw Palace lies on the Al-Faw Peninsula in the far southeast of the Basra province. Water canals from the Shatt al-Arab River turned the land into an agriculturally rich region. Its oil facilities made Al-Faw one of Iraq's major oil exporting ports prior to the 1984 Iran–Iraq war. Because of its strategic and geographic importance, Al-Faw became a target for Iranian control. In February 1986, Iranian units captured the port of Al-Faw. Saddam Hussein vowed to eliminate and remove the enemy "at all cost," and in April 1988, the Iraqi military succeeded in regaining control

The grand entrance to Al-Faw Palace.

of the Al-Faw Peninsula. The palace was built following the 1991 Gulf War in honor of the soldiers who freed the city of Al-Faw from Iranian control. All over the walls inside was written "Victory and glory to the warriors who freed the city from the enemy, the Persians." Now I knew the history of the palace with an ironic name, the "Victory Over Iran Palace."

Chandelier Madness in the Al-Faw Palace

Inside the Al-Faw Palace is a magnificent chandelier. It is as bright as the sun, and its brightness attracts troops and civilian personnel to take photos of it. Standing under the

PART IV: The Fertile Soil of Iraq 355

A magnificent chandelier in the Al-Faw Palace hallway.

spectacular chandelier feels like being on another planet. The Al-Faw Palace hallway looked magnificent and became the headquarters of U.S. coalition forces. It contains 62 rooms and 29 bathrooms. It is situated on a former resort complex, about 3 miles from the IZ. The complex contains several villas and smaller palaces. The interior is evident of the flagrantly extravagant style of the once-dictatorial leader, Saddam Hussein, who made a horrendous mistake and wasted all of the elegance on his self-serving greed.

Lydia and I decided to take individual photos of the chandelier, the long staircase, the huge black marble columns, and the infamous couch that Yasser Arafat gave Saddam Hussein as a gift, which had become one of the troops' favorite spot. If you visit the palace, you might as well sit on the infamous couch. Everything in this building was extraordinary. The palace would soon be turned over to the Iraqis to control their own destiny. As such, the troops and their counterparts flocked to the palace to enjoy the elaborate, grandiose building once used as HQ for the U.S. and Iraqi commanders and the U.S.-led coalition forces. We asked where Saddam Hussein got the marble and the fancy chandeliers. Some said that the marble and fancy chandeliers originated from Italy, whereas others said that they were exported from Spain. Regardless of their origin, the builders used them to make an awe-inspiring architectural tribute to hubris. After the U.S.-led coalition forces withdrew from Iraq, Al-Faw Palace would transform into a new university campus and would become a future modern city.

With my silent freedom, I asked myself several questions: What are the chances of returning to this country 10 or 20 years from now? Would the former war-zone heroes and civilian contractors visit Iraq and relive the memories of a long war? What would Baghdad look like after the rumored modern transformation? Would Babylon also reclaim its fame and restore its vast ancient empire? Most of all, would it be

Lydia and the author inside the marble palace.

safe, and would there be no weapons of mass destruction? How the withdrawal of U.S. troops was managed would help determine future U.S. influence not only in Iraq but in the Middle East as a whole.

The hall was just like a giant football field made of marble. It was humbling to step into the palace, which I had never dreamed of doing in my Army career. The photo-op was more of a payment for family separation and anguish and the indelible sacrifice made in this war-torn country. This palace could have been made into a tool of a positive chain of events that helped improve the lives of Iraqis. Instead, Saddam invaded Kuwait, thus sparking an international crisis. The gruesome attacks on the Twin Towers, the Pentagon, and the plane that went down in a Pennsylvania field will never be forgotten. We forgive but will never forget. The military presence led by the U.S. helped secured and preserve this palace, which is something to remember.

The author with other curious tourists inside Al-Faw Palace.

Every day was special at Al-Faw Palace. It was designed as a corporate retreat and relaxing playground on the lake for members of Saddam Hussein's political party, as a reward for their loyalty and hard work. Now it is a magnet of tourism open to troops and their counterparts as a reward for their loyalty and hard work, where they could spend their short vacations looking around the palace on the man-made lake. The lake served as an excellent way for troops to blow off steam by throwing in a fishing line during their down time. Saddam stocked the lake, and the troops found that there were unique and enormous fish in the waters, known as specially bred "Saddam bass." This was one legacy that American troops found incredibly palatable.

GOING FISHING FOR SADDAM BASS

Saddam stocked the lake with enormous fish.

PART IV: The Fertile Soil of Iraq

On our way to the palace, Lydia and I saw a soldier fishing in the lake. We asked what kind of fish he was trying to catch. He smiled and said he had been trying to catch carp and what he thought were asp, a fish that is a member of the carp family. Other troops call them Saddam bass. He said he was feeling lucky today and knew that he would finally catch a bass. His friends had caught plenty, about 10 or 20, and the bass were really aggressive just before dark. We asked if it would be better if he had his buddies with him to catch more asps, but he said that they were either working out in the gym or sleeping in their chus. Lydia and I left him alone and said we would probably see him again on our way out the palace. With a wide, big grin, he told us to enjoy the last days in the palace. I was glad to hear that the word was out and that everyone was into the mindset of preparing for redeployment.

Unexpected Encounter in the Palace

We missed SGT Crump on this tour, but we saw the 101st Airborne Division's command sergeant major (CSM) and chatted with him. I introduced Lydia, and they got along well. I raised the fact that the CSM cared about his troops and fearlessly visited them for moral support despite the danger during the invasion. He constantly requested logistics support for his units from the commanding general (CG), who approved them without reservation. There were several times when he barely missed RPGs. He was brave like his friend CSM Wilson, who was killed by young terrorists as he and his driver patrolled their regular tour of duty. We were concerned about how it impacted the CSM, but he said he was okay. I said that I thought he was the best there was, and he smiled with modesty.

As we browsed through the palace, I saw another familiar face from the military police (MP) battalion at Camp Henry, Korea. He was the former battalion commander, and it was an unexpected encounter. I arrived in the battalion before he did, and we prepared for his arrival. We had good fun in the battalion, including the time when I helped him with a big party for Hail and Farewell, inviting many high-ranking military guests. It was memorable and a huge success. Because it was an MP Battalion, he gave a safety briefing before the conclusion of the party. The MPs have a unique way of handling things due to their duties. Everyone enjoyed the camaraderie and had fun. The theme was a Hawaiian luau. It was my first big party with a Hawaiian atmosphere. It was very popular, and I knew that people would like it because Hawaii is a giant vacation destination with a tropical

The ground level of the Al-Faw Palace.

First stairway to the offices of U.S.-led coalition forces.

climate, white sand beaches, spectacular waterfalls, and tropical rainforests. It has also become a favorite spot for film making. Colonel Forrest said he remembered that occasion to this day. He was surprised to see me in a dangerous place like Iraq of all places, and we hugged. What a small world! Colonel Forrest is very tall, about 7 feet, and Lydia and I looked like dwarfs in our pictures with him. We could not wait to visit and catch up about the tour in Korea and the future of Al-Faw Palace. He told us to continue with our photo-ops and we would catch up later. Lydia and I tentatively waved goodbye and caught up with the other people touring the palace. The lights were truly fascinating, whether they were real or not, especially the center light. There were American and Iraqi flags hanging side by side on the bannisters. The giant, sophisticated marble columns looked especially solid on the ground floor.

Saddam Hussein's infamous couch inside the Al-Faw Palace.

The author in front of the Al-Faw Palace, Camp Victory.

Bunker number one at the Al-Faw Palace.

Enjoying our time off before troop withdrawal.

Although the Al-Faw Palace was impressive, Lydia and I noticed several places where the marble had come loose, showing chicken wire and plaster underneath. We concluded that there was nothing perfect after all; it was just a big building. Actually, it was really a very tacky building when we looked at it up close. The tour guide told us grisly tales about all of the people murdered by Saddam Hussein during his reign right in the vicinity of the palace. I believed the stories. A few of my old neighbors back in the Philippines had contracts in Iraq, and 95% of them did not come back alive. One of them was my best friend's husband, who was a civil engineer. He applied and took a job in Baghdad so he could provide for his family. But one day, my friend, Virginia, received the terrifying news that her husband had been killed in Baghdad. It was horrifyingly sad news that I would never forget. I also knew a doctor who lived three houses down my street. He successfully landed a contract in Iraq but did not come back alive. One of my brothers-in-law also worked in Iraq as a contractor. He was one of the lucky ones to return home alive. Everything became clear to me about the innocent blood that had been sacrificed in building Saddam Hussein's vainglorious palaces.

My Redeployment

CPT Patrick Franklin was right. There would be heavy traffic on the road as troops withdrew from Iraq. The redeployment of U.S. forces in Iraq, which the Department of Defense refers to as "reposturing," would be an enormous and costly effort. The pieces of equipment in Iraq alone would be worth billions of dollars. The redeployment process from the short war of Operation Desert Storm lasted for about 14 months. A six-year-long battle in Iraq would thus need early planning for the redeployment process. CPT Pat's Corps

PART IV: The Fertile Soil of Iraq

These are some of the Mine Resistant Ambush Protected (MRAP) vehicles that protect U.S. warriors against armor-piercing roadside bombs.

of Engineers unit was among the early units to redeploy, and I missed the opportunity to see him in his convoy from Balad to Camp Striker in Baghdad. I had seen other logistical units preparing to redeploy as well, and I noticed that the drawdown was beginning to affect concessionaires as they gradually closed shop and the soldiers' leisure spots became empty.

As I meditated on my journey to Iraq in support of OIF Parts I, II, III, and IV, I felt proud and fulfilled. I used to think that I had not done anything for the United States. But my mother-in-law, Jane, did not think so. She always reiterated how succesful his son and I were in serving our country. She said she was so proud that both his son and I served the United States with loyalty and selflessness. Our commitment was unquestionable as we continued to serve others the way they supported us during Operation Desert Storm and Operation Iraqi Freedom. She always expressed her undying gratitude and everlasting love whenever she could. Jane's generosity benefitted most of our troops through her ceaseless support in sending support packages to troops. She said that it was the least she could do to support her son, me, and all the men and women who continued to serve this country, the land of the free because of the brave men and women who served and continued to serve.

I love Jane and my in-laws. I sent them flowers and gifts on important occasions, such as birthdays, Mother's Day, Father's Day for Dad, Thanksgiving, Christmas, and just because I wanted to show I care. They are my family. During OIF-I, I sent flowers to Jane on her birthday. I called the flower shop near her residence in St. Louis, Missouri. Jane's favorite flower was the pink carnation. I thought it would look good if I mixed the bouquet with red roses and a pink bow. I could imagine Jane's surprised look as soon as she received her beautiful bouquet. A lady from my favorite flower shop in St. Louis, Missouri took my order and asked for shipping information. She asked what the "From" address should be, so I gave her my address in Iraq. Suddenly the phone went dead. I was afraid I lost her and would have to call the number again and repeat my order. Just before I hung up, the phone came back to life and the lady apologized. I noticed she was quivering and she asked how was it going in Iraq. I told her it was 130°F and was a hard life but we were managing it well. She had been very helpful and said she was going to give me a 20% discount. I thought it was a generous offer. The purchase was completed and a few minutes later, I called Jane and wished her a wonderful birthday. She seemed to always love hearing from me and asked how I was doing in Iraq. I told her everything was manageable and thanked her for her undying love and support for the troops. The support troops received from home was insurmountable and very satisfying. It was this generosity that motivated us to

PART IV: The Fertile Soil of Iraq

give more and accomplish our mission so others can have freedom.

I had packed my duffle bags and carried my backpacks to go to Iraq to work as a soldier and also work alongside soldiers in Iraq so many times. The mission had shifted, and it was now time to withdraw most troops from the battlefield. What were the guarantees that no one would break the peace or if there was even going to be peace, after we left the battlefields of Iraq? Only the future could tell. For the last time as I picked up my backpack to go home and face the unknown, I saw tall blast walls on both sides of the road and saw Iraq in my rearview mirror.

The author signing out.

Afterword

Aurea retired from the 101st Airborne Division, Air Assault (AASLT), and deployed to Iraq in support of Operation Iraqi Freedom for four years: two tours with the 101st and two years as a civilian contractor. She received numerous medals and awards, including two Bronze Star Medals for meritorious service in a combat zone.

The first Bronze Star was received while she was with Bravo Company, 101st Soldier Support Battalion, serving as a Casualty Liaison Team Leader for AASLT in support of Operation Iraqi Freedom in Mosul, Iraq (2003–2004). The second Bronze Star was received for outstanding performance as Human Resources NCOIC, 101st Combat Aviation Brigade, COB Speicher in Tikrit, Iraq (2005–2006).

She holds three master's degrees: two Masters of Business Administration and one Master of Public Administration. She works in Washington, D.C., and continues to support various veterans programs. She is proud to be a lifetime member of the Federal Asian Pacific American Council (FAPAC), previously serving as chairwoman of FAPAC's Mentoring Program and currently serving as Committee Advisor.

www.ingramcontent.com/pod-product-compliance
Lightning Source LLC
Chambersburg PA
CBHW050134240426
43673CB00043B/1661